Aqueous Biphasic Separations

Biomolecules to Metal Ions

Aqueous Biphasic Separations

Biomolecules to Metal Ions

Edited by

Robin D. Rogers

Northern Illinois University
DeKalb, Illinois

and

Mark A. Eiteman

University of Georgia
Athens, Georgia

Springer Science+Business Media, LLC

Library of Congress Cataloging-in-Publication Data

Aqueous biphasic separations : biomolecules to metal ions / edited by
  Robin D. Rogers and Mark A. Eiteman.
      p.   cm.
  Papers presented at a symposium held Mar. 13-14, 1994, at the
207th American Chemical Society National Meeting in San Diego.
  Includes bibliographical references and index.
  ISBN 978-1-4613-5802-2   ISBN 978-1-4615-1953-9 (eBook)
  DOI 10.1007/978-1-4615-1953-9
  1. Phase partition--Congresses. 2. Metal ions--Separation-
-Congresses. 3. Biomolecules--Separation--Congresses.   I. Rogers,
Robin D. II. Eiteman, Mark A. III. American Chemical Society.
Meeting (207th : 1994 : San Diego, Calif.)
QD561.A68   1995
543'.0892--dc20   95-23687
                                                                 CIP

Proceedings of an American Chemical Society Symposium on Aqueous Biphasic
Separations: Biomolecules to Metal Ions,
held March 13–14, 1994, in San Diego, California

ISBN 978-1-4613-5802-2

© 1995 Springer Science+Business Media New York
Originally published by Plenum Press, New York in 1995
Softcover reprint of the hardcover 1st edition 1995

10 9 8 7 6 5 4 3 2 1

PREFACE

Aqueous two-phase systems have been used for the partitioning of cellular particles, biological macromolecules, and smaller organic molecules since the seminal work of Albertsson, who developed the technique in the 1950s. In the last several years, a new group of international scientists has shown such systems also have remarkable utility for the separation of metal ions.

On March 13 and 14, 1994, at the 207th American Chemical Society National Meeting in San Diego, California, the symposium "Aqueous Biphasic Separations: Biomolecules to Metal Ions" was the first such symposium to unite researchers of the two distinct applications of these aqueous partitioning systems. Organizers Robin D. Rogers of Northern Illinois University and Carol K. Hall of North Carolina State University brought together several national and international experts to highlight recent advances in the field and discuss their applications to real world problems in biological and metal ion separations. For many present in the audience, this occasion was their first exposure to the relatively fledgling field of metal ion separations using aqueous biphasic systems.

This book draws on the expertise gathered for the symposium and presents the now fairly diverse work united by the use of aqueous biphasic systems. Chapters include material on metal ion separations, mass transfer effects, affinity partitioning, protein partitioning and refolding, and cell partitioning. Clearly, aqueous biphasic systems are reaching a broader audience. Solvent extraction scientists and engineers are now entering the field, and new lines of communications between the diverse groups are being forged. This symposium showed clearly how each area benefits from the insight and advances brought forth by others.

We hope that this book will not only stimulate the continued communication, but also broaden the general field of aqueous biphasic separations. Our common goal is to harness the incredible potential these systems have for cleaner, cheaper, safer separations of metal ions and for gentle nondenaturing separations of biomolecules.

Robin D. Rogers and Mark A. Eiteman

CONTENTS

METAL ION SEPARATIONS IN POLYETHYLENE GLYCOL-BASED AQUEOUS BIPHASIC SYSTEMS

Robin D. Rogers, Andrew H. Bond, Cary B. Bauer, Jianhua Zhang, Mary L. Jezl, Debra M. Roden, Scott D. Rein, and Richard R. Chomko

Department of Chemistry
Northern Illinois University
DeKalb, IL 60115 USA

INTRODUCTION

Solvent extraction, utilizing an oil/water mixture (e.g., chloroform/water) and a suitable complexant, is a proven technology for the selective removal and recovery of metal ions from aqueous solutions.[1,2] Solving the increasing number of metal ion separation problems has typically focused on finding an appropriate selective extractant, making it as lipophilic as possible, and determining the best diluent. This has often meant the use of expensive extractants and volatile, toxic, organic diluents. In addition, the extracted species is usually dehydrated and partitioned as an ion pair. Whole classes of water soluble extractants cannot be utilized in these systems.

It was somewhat surprising to us, therefore, that aqueous biphasic systems have been virtually ignored as a possible extraction technology for metal ions. (Our recent review of this field revealed a scant nine papers.[3]) Aqueous biphases formed by mixing certain inorganic salts and water soluble polymers, or by mixing two dissimilar water soluble polymers, have been studied for over forty years[4-6] for the gentle, nondenaturing separation of fragile biomolecules. The two aqueous phases (80% water on a molar basis) can be fine tuned to achieve excellent phase separation characteristics (dispersion numbers, densities, and viscosities) often quite similar to widely used oil/water systems.[4,6]

Polyethylene glycol (PEG)-based aqueous biphasic systems have several key advantages over traditional oil/water systems. The PEGs are inexpensive, nontoxic, nonflammable, and commercially available. The salts can be chosen from a variety of inexpensive, relatively harmless possibilities. Toxic, volatile organic diluents are not required and whole new classes of water soluble extractants can be utilized. This opens the door to cleaner, safer, cheaper extraction technologies.

Aqueous biphasic systems also offer unique challenges.[4-6] The number of variables necessary to fully define an optimized process is dramatically increased over traditional systems. The choice of polymer and its molecular weight, the choice of phase forming salt, the relative concentration of each component, the system pH, and the temperature all affect

Aqueous Biphasic Separations: Biomolecules to Metal Ions
Edited by R.D. Rogers and M.A. Eiteman, Plenum Press, New York, 1995

1

the exact phase compositions and therefore the partitioning experiment. In addition, the concentration of salt in the polymer-rich phase and vice-versa, can be quite high leading to excessive losses of the biphase forming components. Finally, reversible partitioning (stripping) is often not straightforward.

We have decided to accept the challenge found in developing aqueous biphasic systems as a proven solvent extraction technology.[7-11] We are attempting to expand the number of phase forming salts studied and thus increase the number of possible solutions we can extract metal ions from. We are also increasing the number of metal ions studied in an attempt to understand the fundamental parameters governing partitioning in aqueous biphasic systems. Perhaps one of the most exciting new horizons this work opens up is the development of new water soluble extractants.[12]

The very nature of PEG-based aqueous biphasic systems makes them ideally suited for several important applications. Extraction may be possible from high ionic strength solutions which themselves salt out PEG. An example of such a "hot" problem pulled from today's headlines is the clean-up of the highly alkaline supernatants in the Westinghouse Hanford waste storage facilities.[13,14] In addition to waste problems, partitioning of metal ion complexes of medicinal importance could allow separation of useful nuclides and direct injection into the human body.[15]

In this chapter we will review our metal ion partitioning work and discuss the three major types of partitioning. These types include: 1) the use of a water soluble extractant which distributes to the PEG-rich phase; 2) the use of halide salts which produce a complex metal anion that partitions to the PEG-rich phase; and 3) those rare instances that the metal ion species present in a given solution partitions to the PEG-rich phase without an extractant. The chapter is divided by metal type, including hard Group 1 and 2 ions, actinides, heavy main group metal ions, and selected transition metals.

EXPERIMENTAL

In the course of preparing a review of metal ion separations in aqueous biphasic systems,[3] we found it difficult to compare distribution results from different works primarily due to the lack of good experimental detail. In aqueous biphasic systems it is well known that there is a dependence of solute distribution on system composition, as well as a number of other factors including temperature and pH.[4-6] Much of the earliest work on metal ion separations using aqueous biphasic systems did not contain this important data on phase compositions, or if compositions were given there was some question as to whether the values were pre- or post-equilibrium numbers. As a result of this ambiguity, comparisons between distribution ratios and distribution characteristics could not be made, and little if any insight into the factors governing metal ion partitioning in these systems could be obtained.

As separations chemists we hope to one day be able to apply some of the novel separations that can be achieved in aqueous biphasic systems. In order to reach this goal an understanding of the factors affecting metal ion distribution behavior must be obtained. The only way to do this is to systematically study the variables and report all pertinent data so that legitimate comparisons between partitioning data can be made. We are therefore taking this opportunity to describe in full detail the approaches taken in our laboratories when studying metal ion distribution in aqueous biphasic systems. It is hoped that this section will serve as a minimum basis for the reporting of pertinent data for distribution studies in these systems.

An important set of variables that should be considered prior to experimentation is the variation in chemical purity of the phase components. This is not typically of major concern for metal salts, water, or the like. The purity of the high molecular weight polymers is, however, an important variable worthy of comment. The polymers, and most notably the PEGs, are used as mixtures of polymers defined by an average molecular weight and the polymer present in highest concentration can vary widely. As a result, partitioning results are dependent upon the PEG used and could differ based on supplier and likely from lot-to-lot from an individual supplier. We are therefore reporting the PEG supplier in the experimental section of our publications and further suggest that researchers check the lot-to-lot consistency of their suppliers. All of the work presented in this chapter has been carried out using PEGs obtained from Aldrich.

Stock solution preparation is carried out on either a weight percent (PEGs or salts) or a molar concentration (salts) basis. We typically prepare several hundred grams of PEG or salt stock solutions and use these throughout a series of uptake experiments. In our publications we always refer to the starting concentration of the stock solutions and generally stress this in the text. The total system compositions can be calculated and the phase compositions taken from the appropriate phase diagram and tie line data. Reporting the stock solution percentages instead of the equilibrium phase compositions allows us to more readily refer to and prepare the systems of interest.

The addition of an extractant to either phase poses a practical problem which we are currently attempting to address. When an extractant is added it is diluted to volume with either the PEG or salt stock solution depending upon solubility. This approach is comparable to that taken by traditional solvent extraction chemists, yet presents difficulties for the aqueous biphase user. By diluting the extractant with stock solution we have maintained a fixed molarity of extractant, but the added mass of the extractant has lowered the weight percent of phase forming component in the PEG or salt stock solution. At low extractant concentrations the effective decrease in salt or PEG weight percent is small, but as the extractant concentration increases the decrease can become significant. Several approaches can be taken to address these difficulties, however, none conveniently parallel the methods of the traditional solvent extraction chemists. Our approach to date has been to carefully state the experimental procedure and to consistently refer to an initial stock solution percentage and the extractant concentration in the phase in which it was prepared.

Further complicating factors are the distribution studies performed at low pH. The addition of acid, usually H_2SO_4 or HNO_3, to these already complex systems introduces yet another experimental variable. Both phases are aqueous hence the pH of both phases, although not necessarily equal, will be affected by adding acid. In an attempt at minimizing the pH gradient between the two phases we have prepared the highly acidic stock solutions for the heavy main group metal studies in the following manner. The PEG or salt is massed and then diluted to the desired weight percent with the acid solution. This affords two stock solutions with known molar concentrations of acid.

The acidic systems utilized in the study of actinide partitioning were prepared in a manner comparable to that taken by traditional solvent extraction chemists. The PEG stock solution was prepared in water while the salt stock solution was diluted to mass with an acid of known concentration.

As an aid to clarity we present an example here. In order to study metal ion uptake at 0.05 M NH_4Br in 2.0 M HNO_3, the following approach was taken. The PEG-2000 and $(NH_4)_2SO_4$ were diluted with a sufficient mass of 2.0 M HNO_3 to afford 40% (w/w) solutions of PEG-2000 and $(NH_4)_2SO_4$ in 2.0 M HNO_3. The salt stock solution was then used to dilute a known mass of NH_4Br to volume to yield a 0.05 M solution. When discussed in the text we refer to, for example, the 0.05 M NH_4Br solution in 40% $(NH_4)_2SO_4$ in 2.0 M HNO_3.

The solutions used in phase diagram determinations were prepared as described above. The binodals of the phase diagrams were determined by turbidity titrations according to the method of Albertsson.[4]

All distribution ratios reported here were carried out using standard radioanalytical methods. Equal volumes (between 750-1000 µL) of the PEG and salt stock solutions were combined and pre-equilibrated by vortexing for 2 min followed by 2 min of centrifugation. The pre-equilibrated biphase system was then spiked with tracer quantities of the nuclide of interest and vortexed for 2 min (a time experimentally determined to be sufficient for the systems discussed here to reach equilibrium), then centrifuged for 2 min to disengage the phases. Each phase was then carefully separated and equal aliquots submitted for standard liquid scintillation or γ-radiometric measurements. The radiometric distribution ratio, D, is defined as the counts per minute in the PEG-rich phase divided by the counts per minute in the salt-rich phase. Radiometric D values are generally accurate to $\pm$ 5%, however, due to the complexity of these systems we report a very conservative accuracy to $\pm$ 10%.

We have carefully explained and outlined our experimental methods in order to stress the importance of defining all of the pertinent variables in common aqueous biphasic partitioning studies. In order to repeat results, make meaningful comparisons between data, and ultimately make progress in developing and testing partitioning theory in these systems, these variables should in some way be defined in each research publication. While our current methodologies are not yet perfected, they provide a suitable compromise between the needs for exact phase compositions and the ability to reproducibly perform numerous distribution ratio experiments. Additionally, the conventions we have adopted are familiar to solvent extraction chemists and may aid in attracting more attention to this field by placing the procedures and results into a more familiar format.

METAL ION PARTITIONING

Group 1 and 2 Metal Ions

There currently exists a need to develop new methods for the selective extraction of Group 1 and 2 metal ions from a variety of media. The presence of Group 1 and 2 cations in certain waste systems has hampered clean-up efforts. One such example is the previously mentioned Westinghouse Hanford tank wastes. In these wastes, high concentrations of ^{90}Sr and ^{137}Cs are present in the highly alkaline supernatants.[13,14] Due to the severe radiological and thermal hazards posed by these nuclides, their removal is extremely important in the overall decontamination of the waste.

Unfortunately, no efficient method for the selective separation of strontium or cesium from such alkaline wastes currently exists. Early work in our group focused on the hydroxide anion which is capable of salting-out PEG-2000 to form a two phase system.[8] We demonstrated that this system is capable of supporting very large concentrations of other inert components (such as extractants, non-phase forming salts, etc.) without destroying the biphasic nature of the system. With this in mind, we set out to test the partitioning behavior of Group 1 and 2 cations under extremely basic conditions.

Our initial studies were aimed at trying to mimic the experimental methods which are currently used in traditional liquid/liquid solvent extraction systems. As a result, we chose to adopt certain conventions when preparing our systems. For instance, as discussed above, when a crown ether is used, it is prepared to a known molarity in a PEG stock solution of a known weight percentage. When matrix ions (such as $NaNO_3$) are present, they are prepared to a known molarity using a salt stock solution of a known concentration. Although these conventions make direct comparisons to traditional oil/water systems possible, they present some complications in describing these complex systems.

In traditional solvent extraction chemistry, it is common practice to investigate the partitioning of a metal ion over a range of extractant concentrations. If we consider the same experiment in an aqueous biphasic system prepared using the above mentioned conventions, increasing the extractant concentration will decrease the overall PEG concentration. Knowing that changing system compositions will affect metal ion distribution ratios, this makes interpretation of any observed trends in distribution ratio vs. extractant concentration very difficult. We had hoped, however, that we would be able to choose extractants that would result in very large distribution ratios for a given metal ion, thus masking the very subtle changes based on system composition effects.

As a starting point, we chose to investigate the use of simple, water soluble crown ethers such as 18-crown-6 and 15-crown-5. This decision was based on several factors including: 1) they are commercially available; 2) they are known to be very selective complexants for Group 1 and 2 cations based on their cavity size; 3) certain variations of these ligands are used very successfully in traditional solvent extraction systems for Sr^{2+} separations;[16,17] and 4) the polyether backbone may facilitate partitioning to the PEG-rich phase.

We chose to investigate the partitioning behavior of Na^+, Rb^+, Cs^+, Ca^{2+}, Sr^{2+}, and Ba^{2+} in a system which was prepared by mixing equal aliquots of a 40% (w/w) PEG-2000 stock solution with a 20% (w/w) NaOH stock solution.[8] In the absence of an added complexant, the distribution ratios for all metal ions in this system are low (ranging from 0.020 for Ca^{2+} to 0.33 for Cs^+). The addition of 18-crown-6 in concentrations ranging from 0.10 M to 1.25 M (concentration in initial PEG stock solution) enhances metal ion partitioning. A distribution ratio above one was achieved for Rb^+.

The addition of relatively large quantities of an extractable anion to boost metal ion distribution ratios is a common practice in traditional systems. In our NaOH/PEG-2000 system, the addition of 2.0 M $NaNO_3$ (concentration in initial salt stock solution) greatly enhances metal ion distribution ratios. All metals (except for Na^+) reach a distribution ratio above unity, with a maximum observed for Ba^{2+} ($D_{Ba} = 7.4$). Although there is an apparent relationship between [18-crown-6] and D_M, no useful information can be obtained from analysis of the slopes of the extractant profiles.

A plot of both D_M and log K (the formation constant for the M^{n+}/18-crown-6 complexes in water) vs. effective ionic radius is presented in Figure 1. A positive correlation is observed for the system with 2.0 M $NaNO_3$ present with a direct correspondence between the log K values and the observed distribution ratios. This would seem to indicate that the principles of metal ion recognition hold for crown ether extraction in aqueous biphasic systems. It is also important to note that the observed maximum distribution ratios are much too large to be explained solely by system composition effects.

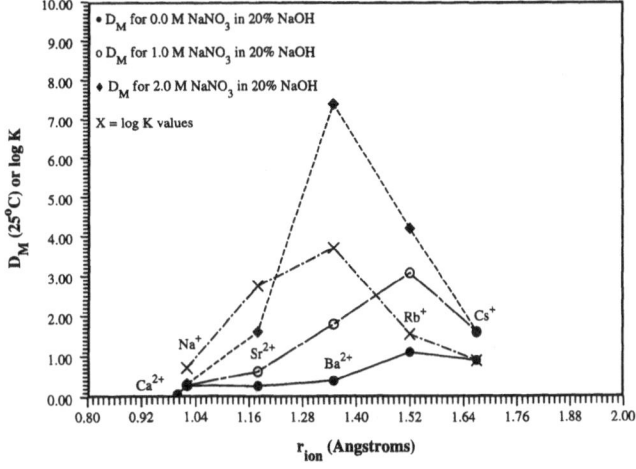

Figure 1. D_M or log K (formation constant for the M^{n+}/18-crown-6 complexes in water) vs. effective ionic radius (X M $NaNO_3$ in 20% NaOH/1.25 M 18-crown-6 in 40% PEG-2000).

The results obtained using 18-crown-6 as an extractant in the NaOH/PEG-2000 system are promising. Unfortunately, gas chromatographic analyses of both phases indicated that the distribution ratio of the crown ether itself is approximately one. An important point to consider in choosing a suitable extractant is that the extractant should partition quantitatively to the phase of interest (the PEG-rich phase here).

It is clear based on the above results that simple crown ether ligands are not the best choice as extractants in these systems. The fact that metal ion recognition in extractant design can be utilized in these systems, however, is particularly satisfying. Due to the complex nature of these systems, a more fundamental approach is needed in order to fully explain the observed results. We are thus continuing our study of Group 1 and 2 metal ion partitioning in aqueous biphasic systems by studying the partitioning behavior of these metals in the absence of any added complexants.

We are exploring the possible correlation of the observed partitioning results with fundamental thermodynamic parameters. Goddard[18] has previously noted that the higher the hydration enthalpy of both cations and anions, the greater the tendency to form a biphase with PEG. Inclusive of this idea is the fact that the higher the hydration enthalpy, the less salt required for biphase formation. It can then be reasoned that the lower the hydration enthalpy, the higher the concentration in the PEG-rich phase and consequently the higher the D_M. We have previously reported a positive correlation between observed distribution ratios and hydration enthalpy for Group 1 and 2 cations.[7] We are expanding these investigations, focusing on the NaOH/PEG-2000 system (over a wide range of system compositions) in order to determine the applicability of using hydration enthalpy data to predict D_M in aqueous biphasic systems. Phase diagrams are particularly useful tools in these investigations. In addition to simply adding to our general knowledge, we are attempting to correlate our phase diagram information for these systems with the observed D_M. This will be particularly useful in determining what effects changing cation and/or anion type will have on D_M.

We are also searching for new water soluble ligands which will act as extractants for Group 1 and 2 cations in these systems. We are attempting to combine the two concepts which are key in successful extractant design. First, the ideas of metal ion recognition will be applied to design ligands which will selectively complex metal ions of interest. Our crown ether work has shown that this is possible in systems of this type. Second, the ligands must be designed so that they report to the PEG-rich phase. This goal is a bit more difficult to accomplish, especially since there is no clear explanation why a neutral ligand should partition in the first place. We are currently using the limited literature in this area, namely that large polyaromatics which are made hydrophilic by sulfonation, will quantitatively partition to the PEG-rich phase.[3] Preliminary results indicate that ligands which are constructed of repeating sulfonated aromatic portions which also contain functionalities capable of complexing metal ions, show greatly enhanced partitioning compared to the crown ether ligands.[12,19]

Actinides

Because of their extreme toxicity, environmental mobility, and inherent radioactivity,[20] a great deal of research has focused on the separation of actinide ions (specifically the transuranic or TRU elements).[21] One problem that hampers the design of new processes to separate these elements is the wide range of waste stream compositions in which the elements are present. For example, at the Westinghouse Hanford site,[14] TRUs exist in solutions ranging from the acidic plutonium finishing plant waste to the highly alkaline single shell tank and complex concentrate wastes. Specialized processes such as TRUEX[14] have been investigated for use with such waste streams, often with great success, however, major modifications such as acidification of the solutions are often required.

The wide range of salts found in these waste streams includes several that can be used to form aqueous biphasic systems. With this in mind we began to investigate the partitioning of the actinides in aqueous biphasic systems. Our initial goal was to build upon the published work,[22-30] as well as to expand the effort by looking at different salt systems and new extractants. As a starting point, we chose to focus on the partitioning behavior of Am^{3+}, Pu^{4+}, Th^{4+}, and UO_2^{2+} because of their relative abundance in many waste systems. For extractants, we chose to investigate the complexing dyes Arsenazo III, Alizarin Complexone, and Xylenol Orange,[9] as well as the water soluble crown ethers 18-crown-6 and 15-crown-5.[10]

In a 40% (w/w) $(NH_4)_2SO_4$/40% (w/w) PEG-2000 system, the distribution ratios for all metal ions investigated are low with values ranging from 0.013 for Am^{3+} to 0.082 for UO_2^{2+}. In this system, Arsenazo III was the best extractant investigated with distribution ratios for Pu^{4+}, Th^{4+}, and UO_2^{2+} near 1000 at 5 x 10^{-3} M Arsenazo III.[9] Alizarin Complexone extracts Th^{4+} and Pu^{4+} well in this system but does not extract the other two metals (D_{Am} and $D_{uranyl} <$ 1). Xylenol Orange will extract all four metals investigated but is best for the tetravalent ions Pu^{4+} and Th^{4+}.

In an alkaline system prepared by mixing 40% K_2CO_3 and 40% PEG-2000, the distribution ratios of all metals are in general much lower compared to the sulfate system regardless of the extractant which is used. Americium is extracted best under these conditions with distribution ratios above ten observed with Xylenol Orange and Alizarin Complexone. Arsenazo III extracts all four metals poorly under these conditions. The general order of extraction in this alkaline system is $Am^{3+} > Pu^{4+} > Th^{4+} > UO_2^{2+}$.

While it is evident that these complexing dyes are capable of achieving very high distribution ratios with the actinide elements investigated, several problems with their use are immediately obvious: 1) the dyes are very hard to work with as they immediately stain all items that they come in contact with including glassware; 2) they are very expensive; and 3) the dyes which achieve the highest distribution ratios (e.g., Arsenazo III) tend to be non-selective. Also, these complexing dyes tend to be toxic, a point which negates a key advantage that aqueous biphasic systems offer. With these factors in mind, we have begun to investigate the use of new ligands as possible actinide extractants.

Our first experiments involved the use of the macrocyclic polyethers 18-crown-6 and 15-crown-5 as extractants in acidic aqueous biphasic systems.[10] The actinide elements (typically uranium, uranyl, and thorium) are known to form hydrogen bonded complexes with crown ethers and the structures of several have been reported.[31-37] Initially we hoped that even if the crown ether ligand did not partition to the PEG-rich phase itself, the formation of a large, hydrogen bonded complex would result in the desired extraction.

In systems prepared by mixing equal aliquots of a 40% PEG-2000 stock solution and a 40% $(NH_4)_2SO_4$ stock solution, the distribution ratios for all metals are less than one (the maximum occurs for UO_2^{2+} with D = 0.085). All metals show enhanced distribution ratios with increasing [18-crown-6]. At 1.25 M 18-crown-6, the distribution ratio for UO_2^{2+} rises above one. The overall order of extraction is $UO_2^{2+} > Pu^{4+} > Th^{4+} > Am^{3+}$. The use of 15-crown-5 as an extractant in this system resulted in lower distribution ratios compared to those observed for 18-crown-6. The exception is Pu^{4+} which shows a slight enhancement with 15-crown-5. The addition of H_2SO_4 or HNO_3 to the salt stock solution results in a depression of the observed distribution ratios as does the addition of NH_4NO_3. This is an interesting trend considering the fact that the addition of the acids and NH_4NO_3 have opposite effects on the phase diagrams. (The addition of acids tends to shift the binodal to the right (higher salt concentrations) while the addition of NH_4NO_3 shifts the binodal to the left.)

If the observed partitioning is due to a hydrogen bonded species with actinide-water-crown ether hydrogen bonds present, the stability of the complex should be related to the stability of the actinide aquo ions. Good correlation has been obtained for the relationship between the observed distribution ratio and the standard enthalpy of formation of the actinide aquo ion. We have also been successful in correlating distribution ratios previously published by Myasoedov[26] with this enthalpic data. A plot of this data is presented in Figure 2. What is not fully clear is whether the observed increase in distribution ratios is due to the formation of a hydrogen bonded species or if it is due to a more basic hydration effect which is independent of crown ether concentration. Most likely it is a complex relationship which may involve a combination of the two.

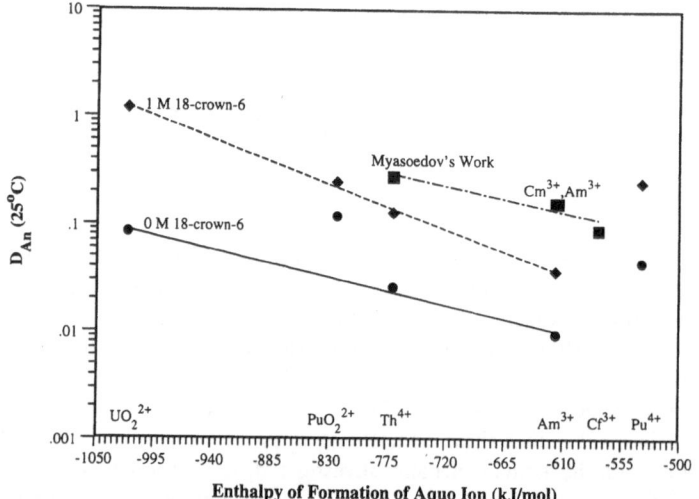

Figure 2. Observed distribution ratios vs. the standard enthalpy of formation of the aquo ions for several actinide elements (40% PEG-2000/40% $(NH_4)_2SO_4$). Myasoedov *et al.*[26] report a total system composition of 15% PEG-2000/14.4% $(NH_4)_2SO_4$ at pH = 4.

From our work it is evident that new water soluble extractants which will complex the actinides and partition to the PEG-rich phase are needed in order to fully exploit these novel systems for the extraction of 5f-metal ions. We are currently investigating the design of such ligands which are less toxic and less expensive than the complexing dyes which have been used previously. In addition, we are continuing our efforts to explain the partitioning results on a much more basic level by examining such areas as thermodynamic data and system effects and how they relate to the observed trends in metal ion partitioning.

Main Group Metal Ions

This group of metal ions, particularly Cd^{2+}, Hg^{2+}, Tl^+, Pb^{2+}, and Bi^{3+}, is comprised of some of the most toxic metals ever to find widespread commercial use. As a consequence, these heavy metals have commonly been found in landfills and environmental restoration sites throughout the world. The well known toxicity of these metals and their complexes, both to man and the environment, has been extensively documented and needs no further discussion here. Adding to these toxicity problems is the presence of these metal ions in sufficient concentrations to adversely affect some separations schemes. For example, some of the Westinghouse Hanford waste tanks contain high concentrations of Bi^{3+} ($BiPO_4$ was used as an actinide precipitant in order to decrease their mobilities), which may interfere with actinide and strontium separations in the TRUEX/SREX combined process.[38]

Much of the early work using aqueous biphasic systems for separations involving this group of metals came from Russian laboratories, and this work has been comprehensively reviewed.[3] More recent work both from Russian[39] and our own[11] laboratories has appeared.

The recent work from the Russian group most notably details a Nb^{5+} purification scheme from PEG-2000/$(NH_4)_2SO_4$ systems containing NH_4F, while our contribution details Bi^{3+} separations with ammonium halides using various PEG-2000/$(NH_4)_2SO_4$ formulations.

The distribution ratios for some of the softer metals in 40% (w/w) PEG-2000/40% (w/w) $(NH_4)_2SO_4$ (initial stock solution percentages) systems with and without 2.0 M acid (2.0 M in both phases) are given in Table 1. All distribution ratios are less than or equal to unity except for Hg^{2+}, which shows enhanced partitioning in acidic media. A full understanding of Hg^{2+} behavior in acidic systems is not yet available and is currently under investigation.

Table 1. Distribution Ratios in the Absence of Complexing Anions in 40% PEG-2000/40% $(NH_4)_2SO_4$ systems at 25°C.

Ion	Solvent	D_M
Cd^{2+}	H_2O	0.066
Cd^{2+}	2.0 M HNO_3	0.15
Cd^{2+}	2.0 M H_2SO_4	0.22
Hg^{2+}	H_2O	0.26
Hg^{2+}	2.0 M HNO_3	10
Hg^{2+}	2.0 M H_2SO_4	12
Tl^+	H_2O	0.12
Tl^+	2.0 M HNO_3	1.0
Tl^+	2.0 M H_2SO_4	1.0
Pb^{2+}	H_2O	0.010
Pb^{2+}	2.0 M HNO_3	0.32
Pb^{2+}	2.0 M H_2SO_4	0.19
Bi^{3+}	H_2O	0.050
Bi^{3+}	2.0 M HNO_3	0.037
Bi^{3+}	2.0 M H_2SO_4	0.043

Nearly all of the aqueous biphasic separations of the late transition and heavy main group metal ions use halide ions as extractants.[3] Extraction of metal halide complexes has been proven to be effective in the usual oil/water extraction systems and is also effective in these polymer/salt systems. Our earliest results in this area stemmed from the unusually high distribution ratios obtained for Bi^{3+} from 1.0 M HCl (D = 2.5) as opposed to 2.0 M H_2SO_4 or HNO_3 (Table 1). A systematic study of the partitioning behavior of Cd^{2+}, Hg^{2+}, Tl^+, Pb^{2+}, and Bi^{3+} as a function of halide ion and its concentration in various matrices was then undertaken.

In contrast to the recently reported Nb^{5+} separations,[39] the lighter ammonium halides in the absence of acid do not effectively transport these softer metals from the lower salt-rich phase to the upper PEG-rich phase. All results from 0.01-1.0 M NH_4F (salt stock solution concentration) in 40% PEG-2000/40% $(NH_4)_2SO_4$ were exceptionally low, following the order $Tl^+ > Hg^{2+} \approx Pb^{2+} > Bi^{3+} \approx Cd^{2+}$ with $D_{max} \approx 0.15$ for Tl^+. All of the extractant dependence profiles are flat and indicate no dependence on $[NH_4F]$.

In order to assess the effectiveness of aqueous biphasic separations at low pH, we have also investigated D_M vs. $[NH_4X]$ (X = F, Cl, Br, I) in 40% PEG-2000/40% $(NH_4)_2SO_4$ systems where both phases are 2.0 M in HNO_3 or H_2SO_4. Because the partitioning results from acid are similar in many cases, they will be discussed immediately after the results without acid for each halide. The $[NH_4F]$ distribution ratio profiles from either 2.0 M HNO_3 or 2.0 M H_2SO_4

are very similar but do differ from the H_2O profiles. The extractant dependencies are all flat, but have spread out (D_{max} for $Hg^{2+} \approx 10$) and have the ordering $Hg^{2+} > Tl^+ > Pb^{2+} > Cd^{2+} > Bi^{3+}$. The high D value for Hg^{2+} can be accounted for by recalling the comparable and unusually high D value obtained from strongly acidic media (Table 1).

Ammonium chloride containing systems provide slightly more enhanced partitioning for these metal ions. Distribution ratios without acid are low and nearly flat from 0.01-0.5 M NH_4Cl for Cd^{2+}, Tl^+, Pb^{2+}, and Bi^{3+}. Hg^{2+} is again the anomaly starting out high, increasing slightly, and actually tailing a bit above 0.5 M NH_4Cl. The D value ordering at 1.0 M NH_4Cl is $Hg^{2+} > Cd^{2+} > Bi^{3+} > Tl^+ > Pb^{2+}$. The addition of acid flattens the extractant profiles for all but Cd^{2+} and Bi^{3+} which increase only slightly. At 1.0 M extractant in strong acid the metal ions partition in the order $Hg^{2+} > Cd^{2+} \approx Bi^{3+} > Tl^+ > Pb^{2+}$, which is comparable to the chloride results in the absence of strong acids.

Use of NH_4Br as the complexing anion markedly enhances the distribution results. Hg^{2+} distribution is relatively constant over the NH_4Br concentration range 0.01-1.0 M, with D values around 750. The partitioning of Cd^{2+} and Bi^{3+} are an order of magnitude lower than Hg^{2+} with maximum D values at 0.5 M NH_4Br near 40 and 20, respectively. Distribution ratios for Tl^+ and Pb^{2+} are at or below one even at the highest [NH_4Br]. In the presence of the strong acids the order of partitioning at the highest [NH_4Br] remains the same as from H_2O only, $Hg^{2+} > Cd^{2+} > Bi^{3+} > Tl^+ > Pb^{2+}$. One important point to note here, and that will be discussed later, is the depression of the D_{Hg} values from a high of around 750 down to 100 in 2.0 M HNO_3, while the distribution of the other metal ions is enhanced in more acidic media. The extractant dependence profiles for Hg^{2+}, Tl^+, and Pb^{2+} are essentially flat and show little dependence on [NH_4Br], while Cd^{2+} and Bi^{3+} D values rise steadily but nonlinearly up to 0.5 M [NH_4Br], where the values obtained from 2.0 M HNO_3 and 2.0 M H_2SO_4 begin to level off.

All of the soft metals we have studied were successfully partitioned from the salt-rich phase above 0.5 M NH_4I in the absence of acid and 0.10 M from both HNO_3 and H_2SO_4 solutions. The order of extraction at 0.5 M NH_4I from H_2O is $Tl^+ > Hg^{2+} > Bi^{3+} > Cd^{2+} > Pb^{2+}$ and is slightly different from the orders obtained with the other halide anions. Pb^{2+} distribution remains low until 0.1 M where it rises sharply; Cd^{2+}, Bi^{3+}, and Tl^+ rise steadily through the intermediate concentrations, and once again the Hg^{2+} profile is flat with the average D value around 800. The NH_4I extractant dependence profiles in 2.0 M HNO_3 and 2.0 M H_2SO_4 are quite similar for all metals with the general order of extraction $Tl^+ > Hg^{2+} > Cd^{2+} > Bi^{3+} > Pb^{2+}$.

Figure 3 shows D_{Cd} vs. [NH_4X] (X = F, Cl, Br, I) in 40% PEG-2000/40% $(NH_4)_2SO_4$ in 2.0 M HNO_3. This profile is typical for the softer metal ions that we have focused on, and nicely illustrates the general trend where the D values decrease in the order $I^- > Br^- > Cl^- > F^-$. This trend holds for systems with and without strong acids. Cd^{2+} has the nicest spread in distribution ratios of the halides; most of the other metal ions have distribution ratios that are similar and relatively low until NH_4I is used. As mentioned earlier, Hg^{2+} is the notable exception. In the absence of strong acids, all of the halide extractant dependence profiles are nearly flat, although a subtle increase is observed for NH_4Cl. The halide ordering in H_2O is more accurately given as $I^- \approx Br^- > Cl^- > F^-$, and in acid as $I^- > Br^- > Cl^- \approx F^-$. Hg^{2+} once again differs in that its distribution ratios for NH_4X (X = Cl, Br, I) are depressed in highly acidic solutions, contrasting the behavior of the other cations where the D values are enhanced at low pH. Figure 4 depicts the usual situation where D_{Cd} from acid is higher than from H_2O only. 2.0 M HNO_3 and 2.0 M H_2SO_4 have comparable effects on the distribution ratios and typically enhance them by about an order of magnitude over those from H_2O only. This effect is reversed for Hg^{2+} where the partitioning results for all but NH_4F are lower in acid. The anomalous behavior for Hg^{2+} is not yet fully understood. Owing to the similarities in the high Hg^{2+} partitioning from the acidic systems (Table 1) and the comparable D_{Hg} for the [NH_4F] dependencies, the effects appear to be real and clearly warrant further investigation.

Figure 5 shows metal ion distribution as a function of [NH_4Br] in 2.0 M HNO_3. Pb^{2+} and Tl^+ distribution ratios are quite low and independent of [NH_4Br], while Hg^{2+} yields the

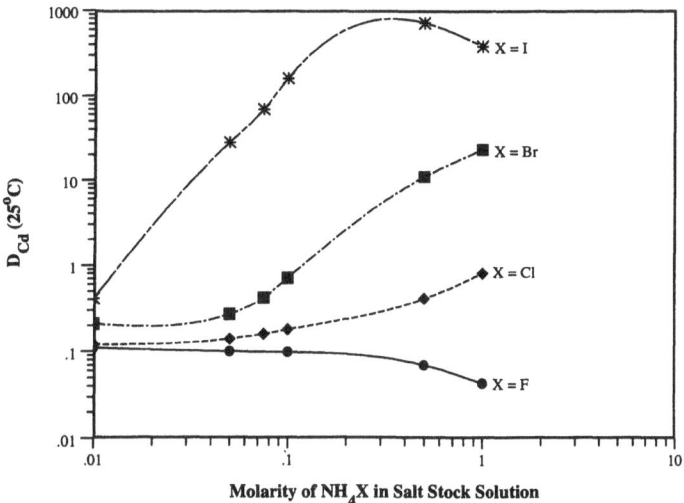

Figure 3. D_{Cd} vs. [NH$_4$X] (X = F, Cl, Br, I) in 40% PEG-2000/40% (NH$_4$)$_2$SO$_4$ in 2.0 M HNO$_3$. This figure illustrates the general trend in increasing distribution ratios as the halide anion is varied from F$^-$ to I$^-$.

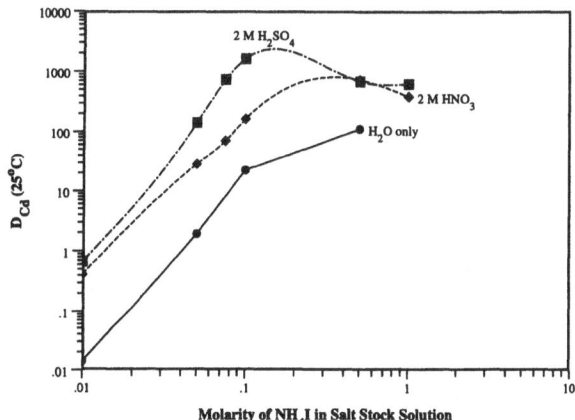

Figure 4. D_{Cd} vs. [NH$_4$I] in 40% PEG-2000/40% (NH$_4$)$_2$SO$_4$. In the halide extraction systems the distribution ratios from highly acidic media are typically higher than those from water alone.

highest D values and is likewise independent of extractant concentration over this range. Cd^{2+} and Bi^{3+} increase slightly and behave very similarly. The order of extraction of these heavy metal ions varies based on halide anion, its concentration, and system pH, but the general extraction sequence can be summarized by observing that Hg^{2+} is typically the best extracted and Pb^{2+} typically the worst. Tl$^+$ is highly dependent upon the halide ion and its concentration, and can increase sharply, but usually remains between the upper and lower limits near Cd^{2+} and Bi^{3+} which behave similarly under a wide range of conditions.

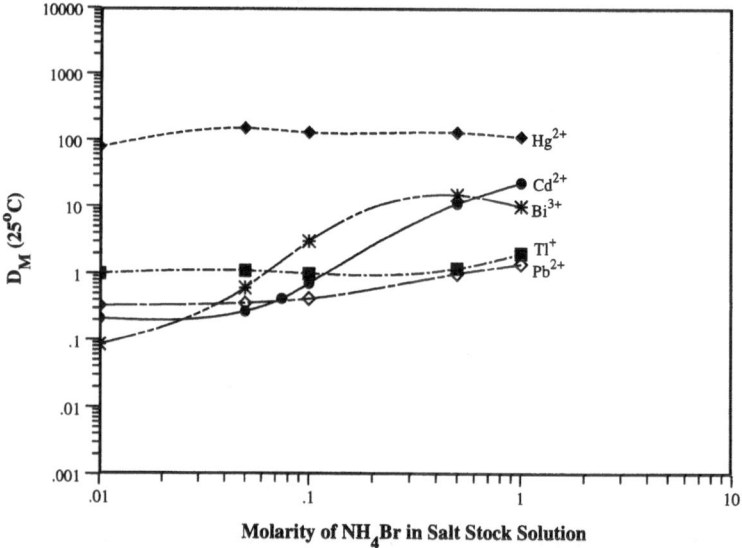

Figure 5. D_M vs. [NH$_4$Br] in 40% PEG-2000/40% (NH$_4$)$_2$SO$_4$ in 2.0 M HNO$_3$. This figure illustrates the general order of extraction of these metal ions, although Tl$^+$ distribution is highly system dependent.

Attempts at explaining the general decrease in metal ion distribution ratios by halides in the order I$^-$ > Br$^-$ > Cl$^-$ > F$^-$ have been carried out. The literature of traditional oil/water separations suggests that in many cases the extracted species is a complex halide anion, and indeed numerous examples exist for the soft cations we are currently studying.[40-50] Interesting correlations have been drawn for D_M vs. log K (formation constants for the complex metal halide anions) as well as D_M vs. charge density of the complex anion. The latter correlation has its roots in the concept that an iodide complex, for example, is highly polarizable and unable to adequately orient its hydration shell and is therefore water destructuring. This type of ion would then be excluded from the highly hydrogen bonded salt-rich phase. These correlations are interesting but not definitive at this point. In a future publication in this area[51] we will focus on gaining a better understanding and explaining the behavior of each metal ion with the different halide anions, determining the partitioning characteristics of the halide extractants, as well as finding new and effective stripping agents to remove the metal halide complexes from the PEG-rich phase.

Pertechnetate, TcO$_4^-$: Partitioning Behavior of Metal Oxyanions

We have discovered that the pertechnetate anion readily partitions quantitatively to the polymer-rich phase of a variety of PEG-based aqueous biphasic systems.[3,15] We were excited by this discovery because the importance of ^{99m}Tc in nuclear medicine and the problems associated with disposal of ^{99}Tc in nuclear waste demand new and better separations technologies for this element.

In radiopharmacy, the short lived ^{99m}Tc ($t_{1/2}$ = 6 h) which decays to ^{99}Tc, is used in the vast majority of all medical procedures utilizing radioisotopes.[52,53] One of the more common ways to access ^{99m}Tc is by eluting pertechnetate from an alumina column containing ^{99}MoO$_4^{2-}$, itself obtained by irradiation of ^{98}Mo or as a ^{235}U fission product. "Instant technetium" involves the solvent extraction of ^{99m}TcO$_4^-$ from an alkaline solution of Na$_2$^{99}MoO$_4$ using methyl ethyl ketone. Both methods suffer disadvantages including the presence of organic impurities and low radiochemical yield.[52-55]

Relatively high levels of $^{99}TcO_4^-$ are present in the highly alkaline waste storage tanks at Westinghouse Hanford[14] and Savannah River.[56] Technetium-99 is a fission product in nuclear fuel burn up. Its long half life (2.12 x 10^5 years) and its environmental mobility (as TcO_4^-) present long term storage problems.[57,58] It is extracted along with other high level waste in the TRUEX and PUREX processes by TBP, and it can interfere with the Pu reduction step by reacting with the reducing agent.

Current extraction technologies for Tc run the gamut from solvent extraction to ion exchange in batch and chromatographic separations, and precipitation reactions.[57] The synthetic organic reagents or resins are often subject to radiation damage (in high level nuclear waste applications) and large cations (e.g., UO_2^{2+}, Zr^{4+}) may be coextracted.[57,59,60] New separations techniques and tailored waste forms are needed for selective removal and immobilization of ^{99}Tc.

Pertechnetate will partition to the polymer-rich phase in PEG-based aqueous biphasic systems from a variety of salt solutions including OH^-, CO_3^{2-}, SO_4^{2-}, and PO_4^{3-} (Figure 6). Increasing the incompatibility between the two phases forces more of the TcO_4^- into the PEG-rich phase. This can be accomplished either by increasing the salt concentration (Figure 6) or increasing the PEG-2000 concentration (Figure 7).

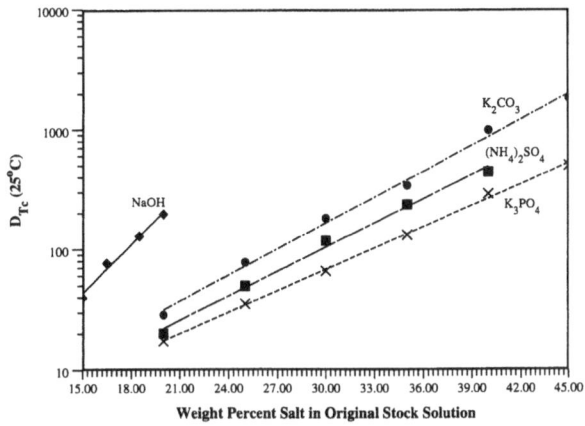

Figure 6. Distribution ratios of TcO_4^- as a function of weight percent of salt stock solution. Each system was prepared by combining 1 mL of 40% PEG-2000 stock solution with 1 mL of salt stock solution.

On a weight percent basis, the trends observed in Figure 6 for D_{Tc} follow the relative abilities of the various inorganic salts to salt out PEG-2000. Here we must decide, however, what criteria define whether one salt is "better" than another at salting out PEG. When considered on a traditional weight percent basis, it takes less NaOH to salt out PEG-2000 and a D value of 200 is achieved at the lowest weight percent by NaOH. On the other hand, when percent salt is converted to molality (Figure 8), the highest D_{Tc} values at lowest salt concentration are observed for the trianion PO_4^{3-}. This is followed by the divalent CO_3^{2-} and SO_4^{2-} salts, and finally by the univalent OH^-. Thus the phospate salt is actually the best of the four salts presented in salting out TcO_4^-. While we will continue to use weight percent for its engineering importance, understanding the processes involved in salting out PEG will require us to also study and report the data in a unit that speaks to the numbers of solute particles present.

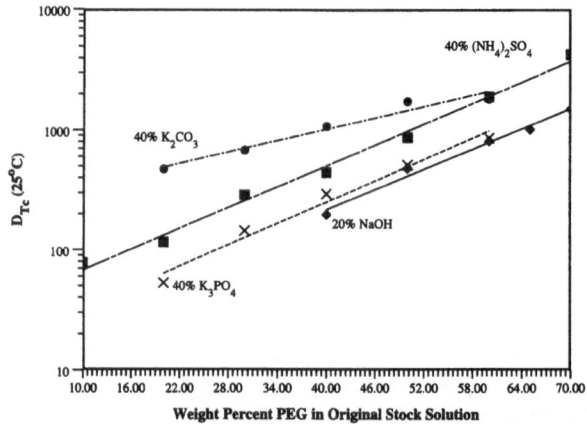

Figure 7. Distribution ratios of TcO$_4^-$ as a function of weight percent of PEG-2000 stock solution.

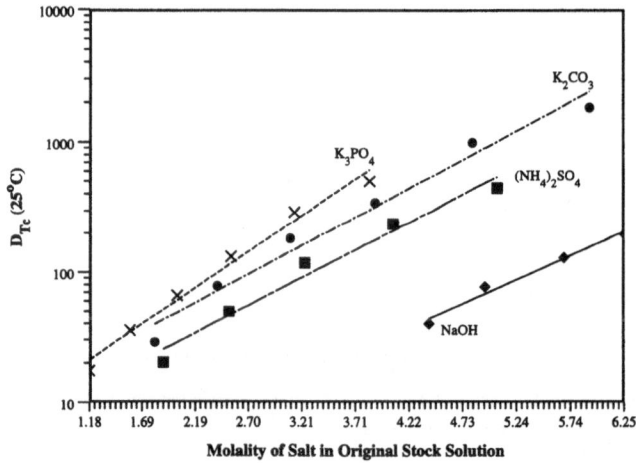

Figure 8. Distribution ratios of TcO$_4^-$ as a function of salt stock solution molality (40% PEG-2000).

The D$_{Tc}$ values of ca. 200 from 4-6 M NaOH raised the possibility of direct extraction of TcO$_4^-$ from the highly alkaline Westinghouse Hanford tank supernatants. Indeed simulants of these waste solutions will form a biphase with 20-70% PEG-2000.[61,62] Preliminary results indicate that simply contacting the waste simulants with 20-70% PEG-2000 will result in TcO$_4^-$ partitioning to the PEG-rich phase. This application of PEG-based aqueous biphasic systems is currently under active investigation.

Another possible application of these systems is in the preparation of high purity [99m]TcO$_4^-$ for medical tracer work. We have investigated the separations of [99]TcO$_4^-$ from solutions of MoO_4^{2-} and have discovered that salts of Group 6 oxoanions, MO_4^{2-}, actually salt out PEG-2000 to form a biphase. Vanadate salts do as well, however, perrhenate and permanganate salts do not.

High distribution ratios for TcO_4^- from these new aqueous biphasic systems reveal quantitative partitioning to the PEG-rich phase (Figure 9). The pertechnetate distribution ratios reveal an interesting anion effect which appears to be related to each anion's charge/size ratio (and hence a variety of properties such as enthalpy of hydration).[18]

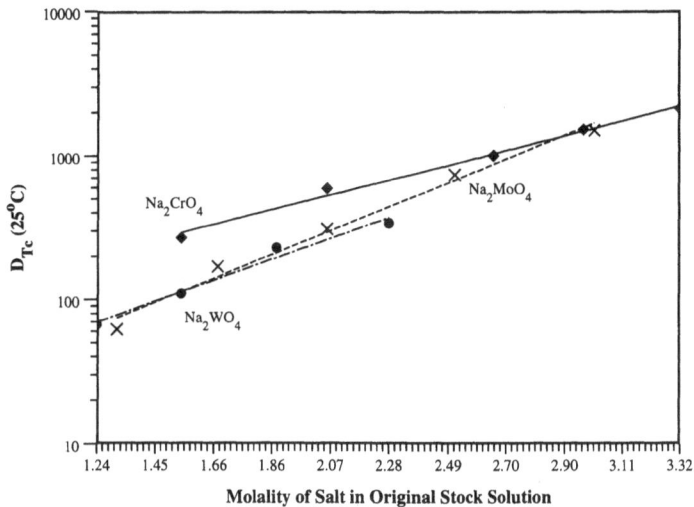

Figure 9. Distribution ratios of TcO_4^- as a function of molality of Na_2MO_4 stock solution (70% PEG-2000).

We have initiated an ambitious project to study TcO_4^- partitioning in PEG-based aqueous biphasic systems as a function of several key parameters including concentration effects, both salt cation and salt anion effects, PEG molecular weight, system pH, and temperature. Thus far the effect on D_{Tc} can be related to the effect on salting out PEG. The linear relationship observed in Figures 6 and 7 for log D_{Tc} versus either weight percent salt or weight percent PEG-2000 appears to be in keeping with the relationship for neutral solutes that suggest that the log of the partition coefficient is proportional to the concentration difference of PEG between the phases.[63] If a cation or anion improves salting out (as they will when they possess a higher charge/size ratio and hence enthalpy of hydration), higher D values for TcO_4^- are observed. Higher molecular weight PEGs, which are salted out with less salt, also exhibit higher D values. In each case studied thus far, the greater the phase incompatibility, the higher the observed value of D_{Tc}.

Perhaps the greatest challenge in finding practical applications for PEG-based aqueous biphasic systems in liquid/liquid extraction lies in finding suitable stripping conditions. Understanding the nature of pertechnetate partitioning behavior has led to an interesting method of stripping. Pertechnetate partitions to the PEG-rich phase because, as a large soft anion, it prefers the hydration environment afforded by the PEG-rich phase rather than the highly ordered water structure in the salt-rich phase.[64] Most metal cations we have studied are highly hydrated and remain primarily in the salt-rich phase. It was anticipated, therefore, that reduction of TcO_4^- to an oxycation form should drive it back to a salt-rich phase, and preliminary data support this idea.

Pertechnetate can be stripped from a loaded PEG-rich phase into a variety of salt stock solutions which contain stannous chloride (as a reducing agent) and a chelating agent to stabilize the reduced species. D_{Tc} values as low as 0.031 have been achieved using a salt stock solution of 0.08 M $SnCl_2$ and 0.06 M citric acid in 40% K_3PO_4 with a pre-equilibrated, pre-loaded PEG-rich phase. Although still very early, one can envisage a process that separates TcO_4^- cleanly from MoO_4^{2-} and then is stripped in reduced form into a solution where it can be directly complexed with an imaging agent and injected into the body.

The stripping scenarios are not as bright when considering waste applications. The necessity of reductive stripping into another salt solution has serious drawbacks when secondary waste streams are considered. This leaves less attractive alternatives such as precipitation or the use of anion exchange resins.

What is clear from our studies of technetium partitioning, is that PEG-based aqueous biphasic systems offer fairly selective media for TcO_4^- removal from a variety of salt solutions. Our current challenge is to understand the fundamental parameters governing these systems so that we can optimize them for specific applications.

CONCLUSIONS

Metal ion separations in aqueous biphasic systems show excellent potential. Separations of cations and anions spanning the periodic table have been demonstrated and discussed. The hard Group 1 and 2 and actinide ions can be extracted into the PEG-rich phase using water soluble complexants and fall into the first aqueous biphase extraction category. By adding water soluble crown ethers to the biphasic system we were able to boost some Group 1 and 2 cation distribution ratios to above one. Further, by correlating distribution ratios with crown ether complex formation constants we showed that metal ion recognition was possible in aqueous biphasic systems. In order to more fully exploit the potential selectivity, water soluble crown ethers that partition exclusively to the PEG-rich phase need to be prepared and evaluated. Additionally, the phase forming salt must be carefully chosen so that it does not compete with the metal ion of interest for crown ether complexation.

Actinide distribution was most readily facilitated by using water soluble complexing dyes like Arsenazo III, Xylenol Orange, and Alizarin Complexone. Separations using these dyes were very effective for certain actinide ions, however, significant difficulties with these separations exist. The use of toxic extractants is clearly unfavorable, the separations were ineffective from basic media (K_2CO_3) due to the deprotonation of the acidic functionalities, and stripping conditions still need to be investigated.

Extraction category two involves the addition of halide anions and the partitioning of anionic complex metal halides. This approach has been proven effective for the soft main group metal ions Cd^{2+}, Hg^{2+}, Tl^+, Pb^{2+}, and Bi^{3+}. Distribution ratios above 1000 were obtained in some systems, even in highly acidic media. A primary drawback to this type of separation is the lack of selectivity afforded by the halide extractant. All metals were effectively partitioned at high NH_4I concentrations, and while this bodes well for potential hard/soft metal separations, the feasibility of intra-group (soft/soft metal) separations is low. Stripping conditions also need to be addressed for these extraction systems.

The third and final type of extraction in aqueous biphasic systems is extraction solely by the PEG-rich phase. This approach was proven successful for pertechnetate anion separations from a wide variety of media. Extraction of TcO_4^- from alkaline (NaOH), basic (K_2CO_3), acidic (($NH_4)_2SO_4$), and highly complex (Westinghouse Hanford tank waste simulants) systems were demonstrated. Stripping is currently one of the major concerns plaguing the two previous extraction categories, however, it has been successfully addressed for pertechnetate separations. By contacting the TcO_4^- loaded PEG-rich phase with a reducing ($SnCl_2$) and chelating (citrate) salt solution, the technetium can be stripped as a cationic species. The numerous medical uses of Tc (as ^{99m}Tc) and its presence in the Westinghouse Hanford waste tanks underscores the importance of TcO_4^- separations. Given the effectiveness and advantages of PEG-based aqueous biphasic separations of pertechnetate, the potential is clear and our group is vigorously exploring this area.

Incredible potential for effective, nontoxic, nonflammable, inexpensive, aqueous-based separations exists. The current drawbacks to aqueous biphasic separations, loss of phase forming components due to solubility in the opposite phase and stripping, can be addressed or are outweighed by the advantages these systems offer. The search for applications of PEG-based aqueous biphasic systems is only beginning and once specific applications are identified these hurdles will be overcome.

In order to realize the full potential of aqueous biphasic separations more research is needed so that an understanding of the parameters governing solute distribution can be obtained. To meet this goal, systematic investigations and clear experimental communication are needed so that the chemistry of the extraction process can be understood. With this knowledge, predictive tools can be developed, new applications of PEG-based aqueous biphasic separations uncovered, and existing technologies enhanced.

ACKNOWLEDGEMENTS

This work was supported by the U.S. National Science Foundation through Grant CTS-9207264. AHB gratefully acknowledges a Laboratory Graduate Assistantship from Argonne National Laboratory.

REFERENCES

1. T. Sekine and Y. Hasegawa. "Solvent Extraction Chemistry. Fundamentals and Applications," Marcel Dekker, New York (1977).
2. "Principles and Practices of Solvent Extraction," J. Rydberg, C. Musikas, and G.R. Choppin, eds., Marcel Dekker, New York (1992).
3. R.D. Rogers, A.H. Bond, and C.B. Bauer, Metal ion separations in polyethylene glycol-based aqueous biphasic systems, *Sep. Sci. Technol.* 28:1091 (1993).
4. P.-Å. Albertsson. "Partition of Cell Particles and Macromolecules," 3rd ed., John Wiley & Sons, New York (1985).
5. "Methods in Enzymology. Aqueous Two-Phase Systems," Vol. 228, H. Walter and G. Johansson, eds., Academic Press, Inc., San Diego, CA (1994).
6. "Partitioning in Aqueous Two-Phase Systems. Theory, Methods, Uses, and Applications to Biotechnology," H. Walter, D.E. Brooks, and D. Fisher, eds., Academic Press, Orlando, FL (1985).
7. R.D. Rogers, C.B. Bauer, and A.H. Bond, Novel polyethylene glycol-based aqueous biphasic systems for the extraction of strontium and cesium, *Sep. Sci. Technol.* in press (1994).
8. R.D. Rogers, A.H. Bond, and C.B. Bauer, The crown ether extraction of group 1 and 2 cations in polyethylene glycol-based aqueous biphasic systems at high alkalinity, *Pure Appl. Chem.* 65:567 (1993).
9. R.D. Rogers, A.H. Bond, and C.B. Bauer, Aqueous biphase systems for liquid/liquid extraction of f-elements utilizing polyethylene glycols, *Sep. Sci. Technol.* 28:139 (1993).
10. R.D. Rogers, C.B. Bauer, and A.H. Bond, Crown ethers as actinide extractants in acidic aqueous biphasic systems: partitioning behavior in solution and crystallographic analyses of the solid state, *J. Alloys Compd.* 213/214:305 (1994).
11. R.D. Rogers, A.H. Bond, and C.B. Bauer, Polyethylene glycol-based aqueous biphasic systems for liquid/liquid extraction of environmentally toxic heavy metals, *in*: "Solvent Extraction in the Process Industries, Proceedings of ISEC'93," Vol. 3, pp. 1641-1648, D.H. Logsdail and M.J. Slater, eds., Elsevier Applied Science, London (1993).
12. C.B. Bauer and R.D. Rogers, The design of metal ion specific extractants for use in polyethylene glycol-based aqueous biphasic systems, paper presented at the 207th American Chemical Society National Meeting, San Diego, CA (1994).

13. M.J. Kupfer, Disposal of Hanford site tank waste, Report WHC-SA-1576-FP, Westinghouse Hanford Company, Richland, WA (1993).

14. W.W. Schulz and E.P. Horwitz, The TRUEX process and the management of liquid TRU waste, *Sep. Sci. Technol.* 23:1191 (1988).

15. D.M. Roden, Y. Song, M.L. Jezl, C.B. Bauer, A.H. Bond, and R.D. Rogers, Pertechnetate partitioning in polyethylene glycol-based aqueous biphasic systems: applications from nuclear medicine to nuclear waste, paper presented at the 207th American Chemical Society National Meeting, San Diego, CA (1994).

16. E.P. Horwitz, M.L. Dietz, and D.E. Fisher, Extraction of strontium from nitric acid solutions using dicyclohexano-18-crown-6 and its derivatives, *Solvent Extr. Ion Exch.* 8:557 (1990).

17. E.P. Horwitz, M.L. Dietz, and D.E. Fisher, SREX: a new process for the extraction and recovery of strontium from acidic nuclear waste streams, *Solvent Extr. Ion Exch.* 9:1 (1991).

18. K.P. Ananthapadmanabhan and E.D. Goddard, Aqueous biphase formation in polyethylene oxide-inorganic salt systems, *Langmuir* 3:25 (1987).

19. R.D. Rogers and C.B. Bauer, unpublished results (1994).

20. "The Chemistry of the Actinide Elements," J.J. Katz, G.T. Seaborg, and L.R. Morss, eds., Chapman and Hall, London (1986).

21. K. L. Nash, A review of the basic chemistry and recent developments in trivalent f-elements separations, *Solvent Extr. Ion Exch.* 11:729 (1993).

22. N.P. Molochnikova, B.F. Frenkel', B.F. Myasoedov, V.M. Shkinev, and Yu.A. Zolotov, Extraction of americium in different oxidation states in a two-phase aqueous system based on poly(ethylene glycol), *Radiokhimiya* 29:39 (1987).

23. N.P. Molochnikova, V.Ya. Frenkel', B.F. Myasoedov, V.M. Shkinev, B.Ya. Spivakov, and Yu.A. Zolotov, Extraction of actinides into aqueous polyethylene glycol solutions from carbonate media in the presence of Alizarin Complexone, *Radiokhimiya* 29:330 (1987).

24. N.P. Molochnikova, V.M. Shkinev, B.Ya. Spivakov, Yu.A. Zolotov, and B.F. Myasoedov, Extraction of actinide and lanthanide complexonates in potassium carbonate-poly(ethylene glycol)-water two-phase aqueous system, *Radiokhimiya* 30:60 (1988).

25. N.P. Molochnikova, V.Ya. Frenkel, and B.F. Myasoedov, Extraction of actinides in two-phase water-poly(ethylene glycol)-salt systems in the presence of potassium phosphotungstate, *J. Radioanal. Nucl. Chem.* 121:409 (1988).

26. B.F. Myasoedov, N.P. Molochnikova, V.M. Shkinev, T.I. Zvarova, B.Ya. Spivakov, and Yu.A. Zolotov, Extraction of complexes of actinides with water-soluble organic reagents in two-phase aqueous systems of salt-poly(ethylene glycol)-water, *in*: "Proceedings of the International Symposium on Actinide/Lanthanide Separations," pp. 164-175, G.R. Choppin, J.D. Navratil, and W.W. Schulz, eds., World Scientific, Singapore (1985).

27. T.I. Nifant'eva, V.M. Shkinev, B.Ya. Spivakov, and Yu.A. Zolotov, Metal extraction in two-phase aqueous systems of the polymer-polymer-salt-water type, *Dokl. Akad. Nauk SSSR* 308:879 (1989).

28. V.M. Shkinev, N.P. Molochnikova, T.I. Zvarova, B.Ya. Spivakov, B.F. Myasoedov, and Yu.A. Zolotov, Extraction of complexes of lanthanides and actinides with Arsenazo III in an ammonium sulfate-poly(ethylene glycol)-water two-phase system, *J. Radioanal. Nucl. Chem.* 88:115 (1985).

29. T.I. Zvarova, V.M. Shkinev, B.Ya. Spivakov, and Yu.A. Zolotov, Liquid extraction in the system aqueous salt solution-aqueous polyethylene glycol solution, *Dokl. Akad. Nauk SSSR* 273:107 (1983).

30. T.I. Zvarova, V.M. Shkinev, G.A. Vorob'eva, B.Ya. Spivakov, and Yu.A. Zolotov, Liquid-liquid extraction in the absence of usual organic solvents: application of two-phase aqueous systems based on a water-soluble polymer, *Mikrochim. Acta* III:449 (1984).

31. G. Bombieri, G. De Paoli, and A. Immirzi, Crown ether complexes of actinide elements. An X-ray study of the conformational change of the crown ether within the uranyl nitrate dihydrate 18-crown-6 molecule, *J. Inorg. Nucl. Chem.* 40:799 (1978).

32. H.S. Du, D.J. Wood, S. Elshani, and C.M. Wai, Separation of thorium from lanthanides by solvent extraction with ionizable crown ethers, *Talanta* 40:173 (1993).

33. P.G. Eller and R.A. Penneman, Synthesis and structure of the 1:1 uranyl nitrate tetrahydrate-18-crown-6 compound, $UO_2(NO_3)_2(H_2O)_2 \cdot 2H_2O \cdot (18\text{-crown-6})$. Noncoordination of uranyl by the crown ether, *Inorg. Chem.* 15:2439 (1976).

34. P.L. Ritger, J.H. Burns, and G. Bombieri, Crystal and molecular structure of $UO_2(NO_3)_2(H_2O) \cdot (12\text{-crown-4})$: correction of the reported structure, *Inorg. Chim. Acta* 77:L217 (1983).

35. R.D. Rogers, L.K. Kurihara, and M.M. Benning, f-Element/crown ether complexes. 10. Oxidation of UCl_4 to $[UO_2Cl_4]^{2-}$ in the presence of crown ethers: structural characterization of crown ether complexed ammonium ions $[(NH_4)(15\text{-crown-5})_2]_2[UO_2Cl_4] \cdot 2CH_3CN$, $[(NH_4)(\text{benzo-15-crown-5})_2]_2[UCl_6] \cdot 4CH_3CN$, and $[(NH_4)(\text{dibenzo-18-crown-6})]_2[UO_2Cl_4] \cdot 2CH_3CN$ and synthesis of $[Na(12\text{-crown-4})_2]_2[UO_2Cl_4] \cdot 2OHMe$ and $[UO_2Cl_2(OH_2)_3] \cdot 18\text{-crown-6} \cdot H_2O \cdot OHMe$, *Inorg. Chem.* 26:4346 (1987).

36. R.D. Rogers, A.H. Bond, W.G. Hipple, A.N. Rollins, and R.F. Henry, Synthesis and structural elucidation of novel uranyl-crown ether compounds isolated from nitric, hydrochloric, sulfuric, and acetic acids, *Inorg. Chem.* 30:2671 (1991).

37. R.D. Rogers, A.H. Bond, and W.G. Hipple, Synthesis and crystal structure of $[UO_2(NO_3)_2(OH_2)_2]\cdot2$(benzo-15-crown-5), *J. Cryst. Spec. Res.* 22:365 (1992).

38. E.P. Horwitz, M.L. Dietz, H.D. Diamond, R.D. Rogers, and R.A. Leonard, Combined TRU-Sr extraction/recovery process, *in*: "Solvent Extraction in the Process Industries, Proceedings of ISEC'93," Vol. 3, pp. 1805-1812, D.H. Logsdail and M.J. Slater, eds., Elsevier Applied Science, London (1993).

39. B.Ya. Spivakov, V.M. Shkinev, L.I. Sklokin, and Yu.A. Zolotov, personal communication (1993).

40. J. Jurkeviciute and M. Malat, Extraction of complex iodide anions with cationogenic tensides. Extraction spectrophotometric determination of divalent mercury, *Chem. Zvesti* 36:91 (1982).

41. M.E. Hofton and D.P. Hubbard, The determination of trace amounts of lead in high-alloy steels by solvent extraction and atomic absorption spectroscopy, *Anal. Chim. Acta* 52:425 (1970).

42. H.A. Mottola and E.B. Sandell, Extraction of bismuth as iodide with isoamyl acetate and isoamyl alcohol, *Anal. Chim. Acta* 24:301 (1961).

43. Yu.M. Yukhin and A.P. Korzhov, Extraction of bismuth from chloride and bromide media by tri-n-butyl phosphate, *Zh. Neorg. Khim.* 22:755 (1977).

44. K. Hasebe and M. Taga, A solvent extraction-spectrophotometric determination of bismuth(III) as tetra-n-butylammonium tetraiodobismuthate, *Talanta* 29:1135 (1982).

45. E.M. Donaldson and M. Wang, Methyl isobutyl ketone extraction of iodide complexes from sulphuric acid-potassium iodide media and back-extraction into an aqueous phase, *Talanta* 33:35 (1986).

46. D.T. Burns and N. Tungkananuruk, Spectrophotometric determination of bismuth after extraction of hexadecyltributylphosphonium tetraiodobismuthate(III) by microcrystalline benzophenone, *Anal. Chim. Acta* 197:285 (1987).

47. D.T. Burns and D. Chimpalee, Spectrophotometric determination of bismuth after extraction of 1-naphthylmethyltriphenylphosphonium tetraiodobismuthate(III), *Anal. Chim. Acta* 211:305 (1988).

48. D.T. Burns, N. Chimpalee, and M. Harriott, Flow-injection extraction spectrophotometric determination of bismuth as tetraiodobismuthate(III) with tetramethylenebis(triphenylphosphonium) cation, *Anal. Chim. Acta* 225:449 (1989).

49. A. Ghosh, K.S. Patel, and R.K. Mishra, Extraction spectrophotometric determination of bismuth(III) with iodide and amidines, *Bull. Chem. Soc. Jpn.* 62:3675 (1989).

50. R.G. Vibhute and S.M. Khopkar, Solvent extraction of antimony(III) with 18-crown-6 from iodide media, *Talanta* 36:957 (1989).

51. R.D. Rogers and A.H. Bond, Polyethylene glycol-based aqueous biphasic partitioning behavior of Cd^{2+}, Hg^{2+}, Tl^+, Pb^{2+}, and Bi^{3+}, *Solvent Extr. Ion Exch.* manuscript in preparation (1994).

52. R.E. Boyd, Molybdenum-99: technetium-99m generator, *Radiochim. Acta* 30:123 (1982).

53. J. Steigman and W.C. Eckelman. "The Chemistry of Technetium in Medicine", National Academy Press, Washington, DC (1992).

54. M.L. Lamson III, A.S. Kirschner, C.E. Hotte, E.L. Lipsitz, and R.D. Ice, Generator-produced $^{99m}TcO_4^-$: carrier free?, *J. Nucl. Med.* 16:639 (1975).

55. A.G.C. Nair, S.K. Das, S.M. Deshmukh, and S. Prakash, Carrier free separation of ^{99}Mo from ^{233}U fission products, *Radiochim. Acta* 57:29 (1992).

56. D.D. Walker, J.P. Bibler, R.M. Wallace, M.A. Ebra, and J.P. Ryan, Jr., Technetium removal processes for soluble defense high-level waste, *Mat. Res. Soc. Symp. Proc.* 44:804 (1985).

57. S. Möbius, Solvent extraction, *in*: "Gmelin Handbook of Inorganic Chemistry. Tc, Technetium: Metal. Alloys. Compounds. Chemistry in Solution," 8th ed., Supplemental Vol. 2, pp. 243-305, H.K. Kugler and C. Kellar, eds., Springer-Verlag, Berlin (1983).

58. C.J. Jones, Applications in the nuclear fuel cycle and radiopharmacy, *in*: "Comprehensive Coordination Chemistry," Vol. 6, pp. 881-1009, G. Wilkinson, R.D. Gillard, and J.A. McCleverty, eds., Pergamon Press, Oxford (1987).

59. T.N. Jassim, J.O. Liljenzin, R. Lundqvist, and G. Persson, Coextraction of uranium and technetium in TBP-systems, *Solvent Extr. Ion Exch.* 2:405 (1984).

60. Z. Kolarik and P. Dressler, Extraction and coextraction of Tc(VII), Zr(IV), Np(IV,VI), Pa(V) and Nb(V) with tributyl phosphate from nitric acid solutions, *Solvent Extr. Ion Exch.* 7:625 (1989).

61. R.D. Rogers, A.H. Bond, C.B. Bauer, Y. Song, J. Zhang, and R.R. Chomko, Polyethylene glycol-based aqueous biphasic systems: inexpensive, nontoxic alternatives for Hanford tank remediation, paper presented at the 208th American Chemical Society National Meeting, Washington, DC (1994).

62. R.D. Rogers, A.H. Bond, C.B. Bauer, J. Zhang, S.D. Rein, R.R. Chomko, and D.M. Roden, Partitioning behavior of ^{99}Tc and ^{129}I from simulated Hanford tank wastes using polyethylene glycol-based aqueous biphasic systems, *Solvent Extr. Ion Exch.* submitted (1994).

63. M.A. Eiteman, Temperature-dependent phase inversion and its effect on partitioning in the poly(ethylene glycol)-ammonium sulfate aqueous two-phase system, *J. Chromatogr. A* 668:13 (1994).
64. R.M. Diamond, The aqueous solution behavior of large univalent ions. A new type of ion-pairing, *J. Phys. Chem.* 67:2513 (1963).

AQUEOUS BIPHASIC SYSTEMS. PROPERTIES AND APPLICATIONS IN BIOSEPARATION

Per-Åke Albertsson

Department of Biochemistry
Chemical Centre
University of Lund
P.O. Box 124
S-221 00 LUND, Sweden

HISTORICAL BACKGROUND

The first experiments on partition of cells and cell particles in aqueous polymer phase systems were carried out in 1955 and published in 1956.[1] These experiments involved the use of one polymer, polyethylene glycol (PEG), together with water and potassium phosphate and the partitioned particles were bacteria, algae, chloroplast fragments, cell walls, and starch grains. It was shown that these particles partitioned either completely to one phase or the other, or they were adsorbed at the liquid-liquid interface. These results agreed with what one would expect from the theories by Brönsted[2] and De Courdes.[3] Later several phase systems with two incompatible polymers were characterized by phase diagrams[4] and used for partition experiments with cell particles[4] and proteins.[5,6] Essentially, these studies showed that proteins could be dissolved and equilibrated between the phases with a reproducible partition coefficient. Separation could be achieved either by a batch procedure or by countercurrent distribution. In the case of particles, the selective adsorption at the interface could also be used for separation in a multistage process such as countercurrent distribution.[7] Thus, cells and cell organelles were analyzed for heterogeneity by countercurrent distribution, a procedure which is reminiscent of liquid-liquid chromatography for soluble small molecules.

In the years 1958-60 the partition behavior of proteins was studied in more detail and it was found that the partition coefficient was a function of both the size of the protein molecules and the ionic composition of the phases. Also, various applications such as concentration of viruses and antigen-antibody binding studies were carried out during this time. This work was summarized in the author's Ph.D. thesis which was published as a book.[8] A second revised edition was published later[9] (also in Russian and Japanese translations) as was a third edition.[10]

Since then a lot has been learned about the mechanism behind partition in aqueous two-phase systems and several applications in bioseparation, both in the research laboratory and on a technical scale, have been published (see below). For a general account of aqueous polymer phase systems in general and various applications, the reader is referred to the books by Walter et al.,[11] Albertsson,[10] and volume 228 of "Methods in Enzymology."[12]

Aqueous Biphasic Separations: Biomolecules to Metal Ions
Edited by R.D. Rogers and M.A. Eiteman, Plenum Press, New York, 1995

21

AQUEOUS POLYMER SYSTEMS

Phase separation in polymer mixtures is a very common phenomenon. In fact miscibility among polymers is an exception rather than the rule. This is due to the high molecular weight of the polymers which results in a relatively low entropy of mixing per unit weight. This in turn means that the interaction between different polymer segments is the dominating factor determining the free energy of mixing and hence whether phase separation occurs or not. If the interaction between segments of two polymers is repulsive, phase separation occurs above certain polymer concentrations and the two polymers will collect separately in two opposite phases. If, however, the interaction between the two polymers is attractive, such as is the case between two oppositely charged polyelectrolytes, the two polymers will collect in the same phase and the result is a phase system with one polymer-rich phase in equilibrium with a polymer-poor phase. The larger the interaction between two polymers the less concentration of polymers is needed to obtain phase separation. To be fully miscible, however, the two polymers must be very similar in their properties.

In Table 1 some pairs of water soluble polymers are listed which have been used in constructing phase systems for use in bioseparation. Figure 1 shows a phase diagram of a dextran-polyethylene glycol (PEG)-H_2O system which has been widely used for separation of various biomolecules such as proteins and nucleic acids and cell organelles such as chloroplasts and mitochondria, membrane vesicles and even whole cells.

Table 1. Some pairs of water soluble polymers which can form aqueous polymer two-phase systems.

Dextran	Polyethylene glycol
Dextran	Methyl cellulose
Dextran	Ficoll
Dextran	Hydroxypropyl dextran
Dextran	Polyethylene glycol
Ficoll	Polyethylene glycol
Hydroxypropyl starch	Polyethylene glycol

The phase systems listed in Table 1 are all due to "incompatibility" between the polymers, i.e., the two polymers collect in separate phases. A particularly interesting combination of two polymers is the pair dextran sulfate-diethylaminoethyldextran (DEAE-dextran).[10] At low ionic strength, and at neutral pH, the interaction between these two polymers is attractive because dextran sulfate is negatively charged and DEAE-dextran is positively charged. The two polymers will therefore collect in the same phase. At high ionic strength this attraction is weakened and the hydrophobic diethyl groups of DEAE-dextran will dominate the interaction between the two polymers and make it repulsive such that the two polymers will be incompatible and collect in opposite phases. At certain intermediate ionic strengths the attractive and repulsive forces will cancel and the two polymers are miscible. This phase system, therefore, displays all the three phenomena which are possible when mixing two polymers and should be an interesting system for theoretical studies of phase separation in polymer mixtures.

A two-phase system can be described by a phase diagram where the composition of the two conjugated phases in equilibrium are given for various mixtures of the two polymers in water (Figure 1). Phase diagrams of several other systems can be found in reference 10. Aqueous polymer-polymer two phase systems are characterized by the following properties. The difference in density as well as in refractive index between the phases is small. The interfacial tension between the phases is small, 0.1-100 N/m. The settling time (phase separation at normal gravity) is relatively long (15 min-several hours) due to the small density difference and the relatively high viscosity of the phases. The time for separation can be reduced, however, by low speed centrifugation.

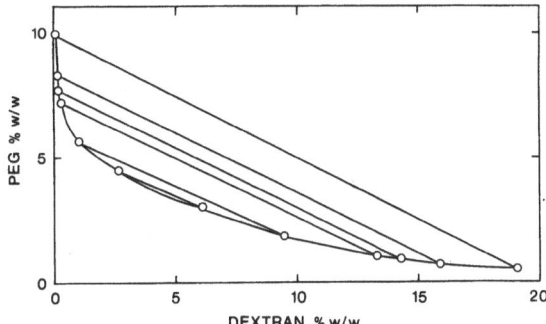

Figure 1. Phase diagram of a Dextran-polyethylene glycol-water phase system at 20°C (Dextran 500 (Pharmacia) molecular weight 500,000 (weight average), polyethylene glycol (Carbowax PEG 8000), molecular weight 8,000). Dextran is enriched in the lower phase and polyethylene glycol in the upper phase.[10]

One can construct isopycnic phase systems (e.g., dextran-Ficoll-H_2O) with no density difference.[10] In such a phase system, one phase will eventually form a globular phase immersed in the other, like egg yolk in an egg. The behavior is similar to the behavior of other phase systems at zero gravity, which has been studied in one of the space shuttles.[13]

Three phase systems are obtained if solutions of three different polymers, all mutually incompatible, are mixed. With four polymers a four phase system is obtained, etc.[10] Many polymers form two-phase systems when combined with suitable salts (e.g., phosphates, sulfates, citrates). These systems have found use for technical extraction of enzymes.[14]

An aqueous polymer two-phase system can also be obtained with only one polymer and water. For example, water solutions of random copolymers of polyethylene glycol and polypropylene glycol form a two phase system above certain temperatures and such systems have also been used in bio-separation.[15]

FACTORS DETERMINING PARTITION

Partition depends on many factors originating from both of the phase polymers, the ionic composition, and the partitioned substance. Among the phase polymer factors are type of polymers and their molecular weights, as well as presence of certain chemical groups on the polymer (e.g., ionized, hydrophobic, or biospecific groups). The ionic composition is of paramount importance because the ions present in the system determine the sign and magnitude of the interfacial electric potential.[10] Among the properties of the partitioned substance that determine its partition are size, charge, and hydrophobic properties of the surface, and presence of receptors for biospecific ligands (attached to one of the phase polymers). Also, the chirality of molecules can be reflected in the partition.[16] Generally small molecules partition evenly between the phases unless there is a binding molecule in one of the phases.

The different factors that determine partition can be explored separately or in combination to achieve an effective separation. Some of the factors can be enlarged so that they will dominate the partition behavior. With respect to the partitioned substances, the following types of partition can be distinguished.

1. Size-dependent partition. Molecular size or the surface area of the particles (or molecules) is the dominating factor.

2. Electrochemical partition. Electrical potential between the phases is used to separate molecules or particles according to their charge.

3. Hydrophobic affinity partition. Hydrophobic properties of a phase system is used for separation according to the hydrophobicity of molecules or particles.

4. Biospecific affinity partition. The affinity between sites on the molecules or particles and affinity ligands attached to one of the phase polymers is used for separation.

5. Conformation-dependent partition. Conformation of the molecules and particles is the determining factor.

6. Chiral partition. Enantiomeric forms are separated.

The partition of a solute (e.g., a protein) between the phases is described by a partition coefficient, K, defined as the ratio between the concentrations of solute in the upper and lower phase. Formally, we can resolve the partition coefficient into a number of factors:

$$K = K^o K_{el} K_{hfob} K_{biosp} K_{size} K_{conf} \tag{1}$$

where the indices *el, hfob, biosp, size,* and *conf* stand for electrochemical, hydrophobic, size-dependent, and conformational contributions respectively, to the partition coefficient. The K^o part includes all other factors, such as general relative solvation of the solute molecule in the phases. The logarithmic form of the relation above is especially useful when the various effects are studied:

$$\log K = \log K^o + \log K_{el} + \log K_{hfob} + \log K_{biosp} + \log K_{size} + \log K_{conf} \tag{2}$$

Effects of a number of parameters on the partition of proteins and nucleic acids are collected below.

Polymer Concentration

Increasing the polymer concentration moves the phase system further from the critical point and the two phases become more different. As a result, soluble substances such as proteins partition more toward one of the phases and often the separation factor (i.e., the ratio between the partition coefficients of two components) increases. The viscosity of the phases increases with increasing polymer concentration which one has to consider in finding an optimal phase system.

Particles such as cell organelles or membrane particles usually adsorb more strongly to the interface when the polymer concentration is increased. This adsorption to the interface is selective and can be utilized for separation purposes. For example, plasma membranes can be purified from other cellular membranes by three repeated batch extractions in a dextran/polyethylene glycol phase system.[17]

Molecular Weight of Polymers

A general rule holds for the effect of the molecular weight of polymers on partition. Partitioned molecules or particles will favor a phase more if the molecular weight of its polymer is reduced. For example, the partition coefficient of a protein in the dextran/polyethylene glycol system will increase if the molecular weight of polyethylene glycol is decreased or the molecular weight of dextran is increased. Conversely, the partition coefficient of the protein will decrease if the molecular weight of polyethylene glycol is increased or the molecular weight of the dextran is decreased.

This effect of the molecular weight of polymers in turn depends on the molecular weight of the partitioned protein. Proteins with higher molecular weights are more influenced by changes in the molecular weight of polymers than proteins with small molecular weights. Polymers with different molecular weights can therefore be used to optimize separation of protein molecules differing in size.[18]

Electrochemical Potential

The salt composition is of paramount importance in partitioning of all kinds of molecules and cell particles. Although salts partition fairly evenly between the phases, there are small but significant differences in the partition coefficient of different salts,[19,20] meaning that different ions have different affinities for the two phases. An electrical potential between the phases is therefore created.[9,10] For a salt, the ions of which have the charges Z^+ and Z^-, the interfacial potential ψ is given by:

$$\psi = [RT/(Z^+ + Z^-)F] \ln (K_-/K_+) \tag{3}$$

where R is the gas constant, F the Faraday constant, T the absolute temperature, and K_- and K_+ hypothetical partition coefficients of the ions in the absence of a potential. The interfacial potential will be larger the larger the K_-/K_+ ratio. A salt with two ions that have very different affinities for the two phases will generate a larger potential difference than a salt with ions that have similar affinities for the phases.

It can be shown that, in the presence of excess of salt, a protein will partition according to:

$$\ln K_p = K_p^o + (FZ/RT)\psi \tag{4}$$

where K_p is the partition coefficient of the protein and K_p^o is the value of this coefficient when the interfacial potential (generated by the excess salt) is zero or when the protein net charge Z is zero. Thus the difference in affinity of the ions for the two phases generates an electric potential difference according to equation (3), which in turn affects the partition of the protein according to equation (4). Even if ψ is small, it will strongly influence K_p because Z is a large number for most proteins. K_p changes exponentially with Z. Experimental results agree very well with equation 4. As shown, for example, in Figure 2, the log K for a protein is linearly dependent on the net charge as expected from equation 4.[21] The slope of the lines in Figure 2 is a measure of the electrochemical potential, ψ, between the phases. This in turn is dependent on the type of salt which is in excess in the phase system (equation 3). An experiment like that of Figure 2 is the most straightforward and reliable way of determining the electrochemical potential. Using electrodes in the system for measuring the interfacial electrical potential, which has been tried by others,[22,23] will only introduce additional factors and problems such as liquid junctions, electrode surface effects, etc. For a recent discussion of the electrochemical potential between the phases see reference 24.

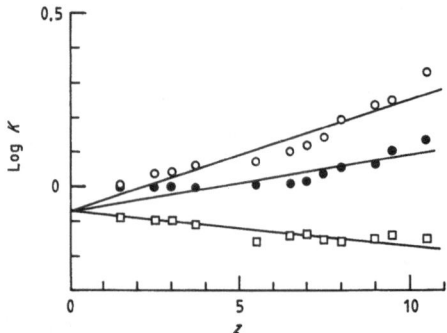

Figure 2. Relationship between partition coefficient, K, and net charge of ribonuclease in an aqueous two phase system of dextran and polyethylene glycol with three different salts: 0.1 M KSCN (o), 0.1 M KCl (●) or 0.05 M K_2SO_4 (□). The slopes of the lines are proportional to the interfacial electrical potential, see equation 4 in text.[21]

The influence of different salts, and hence the electrochemical potential, on the partition of proteins in the dextran/polyethylene glycol system has been studied in detail.[10,20,21] It is the ratio between different ions which is the most important factor, while the ionic strength *per se* is of minor influence. The cations decrease the partition coefficient of negatively charged proteins in the order $Li^+ < NH_4^+ < Na^+ < Cs^+ < K^+$ and the monovalent anions in the order $F^- < Cl^- < Br^- < I^-$. The divalent anions HPO_4^{2-}, SO_4^{2-}, and citrate^{2-}, increase the partition coefficient relative to the monovalent anions. For positively charged proteins the orders given above are reversed. In summary, to achieve a high partition coefficient for negatively charged proteins, one should use lithium or sodium monohydrogen phosphate, sulfate, or citrate. On the other hand, to get a lower partition coefficient, one should use potassium chloride or bromide. For a positively charged protein, the opposite will hold. The strong effect of the ionic composition on the partition holds for all charged macromolecules and particularly so for cell particles which each carry a vast number of charges.

Cross Partition

The isoelectric point of proteins can be determined by cross partition.[10] Using this method the partition coefficient is determined as a function of pH with two different salts (in excess over buffer). Two curves are obtained, one for each salt, when the partition coefficient is plotted versus pH. The two curves cross at the isoelectric point. This method can also be used for the determination of the isoelectric point of cell organelles, membrane vesicles, and cells.

Charged Polymers

A charged polymer, concentrated in one of the phases, shows a stronger effect on the partition of charged biomacromolecules than salts do. However, the concentration of low-molecular electrolytes in the system must be kept low compared with the concentration of polymer-bound charged groups. Charged polyethylene glycol (PEG) has been used for directing the partition of proteins. Positively charged trimethylamino-PEG (TMA-PEG) extracts negatively charged proteins into the PEG-rich upper phase while positively charged proteins are excluded. Negatively charged PEG (e.g., PEG sulfonate (S-PEG) or carboxymethyl-PEG (CM-PEG) has the opposite effect. By changing the pH value in the system, the net charge of proteins present can be successively varied and their partition coefficients adjusted within extreme values.[25,26]

Hydrophobic Groups

By introducing a low concentration (1 mM or less) of PEG-bound hydrophobic groups (e.g., palmitate), proteins with hydrophobic binding sites show increased affinity for the upper phase.[26,27] Also membrane vesicles and cells can be extracted into the PEG-phase even at very low concentrations of the PEG-palmitate.[28]

Affinity Ligands

Affinity ligands attached to one of the polymers have been used to extract ligand-binding proteins and nucleic acids into the corresponding phase. The use of a triazine dye as affinity ligand, is particularly useful.[29] The log K_{biosp} is obtained as the difference in the log K values observed with excess of ligand-PEG and with no ligand present. The efficiency in extraction measured in this way increases with the concentration of phase-forming polymers.

Temperature

The composition of the phases changes with the temperature. Proteins partition more equally between the phases of a two-polymer systems, usually, when the temperature is increased. This effect can be counteracted by using higher concentrations of the polymers.

APPLICATIONS

Phase partition using aqueous polymer systems has been applied for separation of a large number of different molecules and particles of biological origin. The partition method is very versatile. It can be applied on a small scale (down to microliters) as well as on a large industrial scale (volumes of cubic meters).

Batch procedures are applied for preparative purposes. These involve a few extraction steps either with a constant phase composition or by a "gradient" procedure where the composition of the phase system is changed after each partition step in order to sequentially extract the different components of a mixture.[10,30]

Multistage procedures, such as countercurrent distribution or partition chromatography, are used for analytical purposes when one is interested in the study of the heterogeneity or complexity of a mixture. By these procedures one may also discover new molecules or particles. Some examples of bioseparation which illustrate the versatility of the partition method follow here.

Proteins. Several enzymes have been purified by batch procedures using affinity partition where a specific ligand was bound to one of the polymers.[29] The salt-PEG phase system has been widely used for large scale purification of enzymes from a crude lysate of bacteria and cell homogenates.[14]

Of particular interest is the application of partition chromatography. One phase, the dextran-rich phase of the dextran-PEG system, is bound to a solid support while the other, the PEG-rich phase, is mobile.[31,32] This has recently been used for the analysis of immunoglobins and antigen-antibody complexes.[33]

Nucleic acids. These have been analyzed by partition chromatography and can be separated both according to size and base composition.[31,32]

Cell organelles and membrane vesicles. Chloroplasts and mitochondria have been purified by either batch or countercurrent distribution procedures.[34,35] Plasma membranes, from both animal and plant cells, have been purified by batch procedures involving three partition steps.[35,36] Specific lectins bound to the dextran in the bottom phase of the dextran-PEG system have been used to withdraw plasma membranes (which expose sialic acid groups on their outer surface) from other cellular membranes which, with a selected ionic composition, partition into the upper PEG-phase. By such biospecific affinity partition, very selective and efficient purification can be obtained rapidly with only one or two partition steps.[37]

Of particular interest is the combination of fragmentation with separation in order to analyze the structure of biological membranes.[38,39] The photosynthetic membrane of chloroplasts has been fragmented by sonication followed by countercurrent distribution to separate the different membrane vesicles so obtained. Biochemical analysis of the different fragments and comparison with theoretical models has allowed the construction of realistic models for the structure and function of the photosynthetic membrane.

Viruses. Different viruses have been enriched and concentrated from a solution by adding polymers so that a phase system with, e.g., a very small bottom phase into which the virus partitions, is formed.[40] This is a very mild and efficient way of concentrating viruses from large volumes of culture media. This technique should be applicable to other particles too, since it is a general phenomenon that particles often partition one-sidedly into one of the phases.

Cells. Various cells such as erythrocytes, leucocytes, cultured tissue cells, algae, and bacteria have been partitioned, usually with the dextran-PEG phase system. The partition depends on the surface properties of the cells and these can, e.g., be separated according to their stage in the cell's life cycle.[12] Some mutants of bacteria show different partition behavior due to changes in the composition of their surface layer. Also the isoelectric points and the hydrophobicity of the bacterial surface has been determined and compared with the virulence of pathogenic bacteria.[41]

CONCLUSION

A large number of data has now been collected concerning the behavior of biomacromolecules and cell particles. These show that reproducible partition can be obtained if conditions such as temperature and phase composition are carefully controlled. We also know most of the factors which determine partition. Future research should be directed to find new phase systems with increased selectivity and polymers which allow an economical application on a technical scale.

REFERENCES

1. P.-Å Albertsson, Chromatography and partition of cells and cell fragments, *Nature* 177:771 (1956).
2. J.N. Brönsted, Molekülgrösse und phasenverteilung. I, *Z. Phys. Chem., Abt. A* Bodenstein-Festband:257 (1931).
3. T. De Courdes, *L. Rhumbler, Wilhelm Roux' Arch. Entwicklungsmech. Org.* 7:225 (1898).
4. P.-Å. Albertsson, Particle fractionation in liquid two-phase systems. The composition of some phase systems and the behavior of some model particles in them. Application to the isolation of cell walls for microorganisms, *Biochim. Biophys. Acta* 27:378 (1958).
5. P.-Å. Albertsson, Partition of proteins in liquid polymer-polymer two-phase systems, *Nature* 182:709 (1958).
6. P.-Å. Albertsson and E.J. Nyns, Counter-current distribution of proteins in aqueous plumer phase systems, *Nature* 184:1465 (1959).
7. P.-Å. Albertsson and G. Baird, Counter-current distribution of cells, *Exp. Cell Res.* 28:296 (1962).
8. P.-Å. Albertsson. "Partition of Cell Particles and Macromolecules," Almqvist & Wiksell, Stockholm; Wiley, New York (1960).
9. P.-Å. Albertsson. "Partition of Cell Particles and Macromolecules," 2nd ed., Almqvist & Wiksell, Stockholm; Wiley (Interscience), New York (1971).
10. P.-Å. Albertsson. "Partition of Cell Particles and Macromolecules," 3rd ed., Wiley, New York (1986).
11. "Partitioning in Aqueous Two-Phase Systems. Theory, Methods, Uses, and Applications to Biotechnology," H. Walter, D.E. Brooks, and D. Fisher, eds., Academic Press, Orlando, FL (1985).
12. "Methods in Enzymology. Aqueous Two-Phase Systems," Vol. 228, H. Walter and G. Johansson, eds., Academic Press, Inc., San Diego, CA (1994).
13. J.M. Van Alstine, J. Boyce, M. Harris, S. Bamberger, P.A. Curreri, R.S. Snyder, and D.E. Brooks, Interfacial factors affecting the demixing of aqueous polymer two-phase systems in microgravity, *in*: "Proceedings of an NSF Workshop on Interfacial Phenomena," Space Science Laboratory preprint series No. 86-139 (NASA) (1986).
14. M.-R. Kula, K.H. Kroner, and H. Hustedt, Purification of enzymes by liquid-liquid extraction, *in*: "Advances in Biochemical Engineering," Vol. 24, pp. 73-118, A. Fiechter, ed., Springer Verlag, Berlin (1982).
15. P.A. Alred, A. Kozlowski, J.M. Harris, and F. Tjerneld, Application of temperature-induced phase partitioning at ambient temperature for enzyme purification, *J. Chromatogr.* 659:289 (1994).
16. B. Ekberg, B. Sellergren, and P.-Å. Albertsson, Direct chiral resolution in an aqueous two-phase system using the counter-current distribution principle, *J. Chromatogr.* 333:211 (1985).
17. P. Kjellbom and C. Larsson, Preparation and polypeptide composition of chlorophyll-free plasma membranes from leaves of light-grown spinach and barley, *Physiol. Plant.* 62:501 (1984).
18. P.-Å. Albertsson, A. Cajarville, D.E. Brooks, and F. Tjerneld, Partition of proteins in aqueous polymer two-phase systems and the effect of molecular weight of the polymer, *Biochem. Biophys. Acta* 926:87 (1987).
19. G. Johansson, Partition of salts and their effects on partition of proteins in a dextran-poly(ethylene glycol) water two-phase system, *Biochim. Biophys. Acta* 221:387 (1970).
20. G. Johansson, Effects of salts on the partition of proteins in aqueous polymeric two-phase systems, *Acta Chem. Scand. B* 28:873 (1974).
21. G. Johansson, Partition of proteins and micro-organisms in aqueous biphasic systems, *Mol. Cell Biochem.* 4:169 (1974).
22. D.E. Brooks, K.A. Sharp, S. Bamberger, C.H. Tamblyn, G.V.F. Seaman, and H. Walter, Electrostatic and electrokinetic potentials in two polymer aqueous phase systems, *J. Colloid Interf. Sci.* 102:1 (1984).
23. R. Reitherman, S.D. Flanagan, and S.H. Barondes, Electromotive phenomena in partition of erythrocytes in aqueous polymer two phase systems, *Biochem. Biophys. Acta* 297:193 (1973).
24. C.A. Haynes, J. Carson, H.W. Blanch, and J.M. Pausnitz, Electrostatic potentials and protein partitioning in aqueous two-phase systems, *AIChE J.* 37:1401 (1991).
25. G. Johansson, A. Hartman, and P.-Å. Albertsson, Partition of proteins in two-phase systems containing charged poly(ethylene glycol), *Eur. J. Biochem.* 33:379 (1973).
26. G. Johansson, Uses of poly(ethylene glycol) with charged or hydrophobic groups, *in*: "Methods in Enzymology. Aqueous Two-Phase Systems," Vol. 228, pp. 64-74, H. Walter and G. Johansson, eds., Academic Press, Inc., San Diego, CA (1994).
27. V.P Shanbhag and V.P. Axelsson, Hydrofobic interaction determined by partition in aqueous two-phase systems. Partition of proteins in systems containing fatty acid esters of poly(ethylene glycol), *Eur. J. Biochem.* 60:17 (1975).
28. E. Eriksson, P.-Å. Albertsson, and G. Johansson, Hydrofobic surface properties of erythrocytes studied by affinity partition in aqueous two-phase systems, *Mol. Cell. Biochem.* 10:123 (1976).

29. G. Kopperschläger, Affinity extraction with dye ligands, *in*: "Methods in Enzymology. Aqueous Two-Phase Systems," Vol. 228, pp. 121-136, H. Walter and G. Johansson, eds., Academic Press, Inc., San Diego, CA (1994).

30. T.G. Hammond, R.R. Majewski, J.J. Onorato, P.C. Brazy, and D.J. Morré, Isolation and characterization of renal cortical membranes using an aqueous two-phase partition technique, *J. Biochem.* 292:743 (1993).

31. W. Müller, Columns using aqueous two-phase systems, *in*: "Methods in Enzymology. Aqueous Two-Phase Systems," Vol. 228, pp. 100-112, H. Walter and G. Johansson, eds., Academic Press, Inc., San Diego, CA (1994).

32. W. Müller, Separation of proteins and nucleic acids, *in*: "Methods in Enzymology. Aqueous Two-Phase Systems," Vol. 228, pp. 193-206, H. Walter and G. Johansson, eds., Academic Press, Inc., San Diego, CA (1994).

33. K. Andersson, C. Wingren, and U.-B. Hansson, Liquid-liquid partition chromatography as a method to examine surface properties of antibodies and antigen-antibody complexes, *Scand. J. Immunol.* 38:95 (1993).

34. M.J. López-Pérez, Preparation of synaptosomes and mitochondria from mammalian brain, *in*: "Methods in Enzymology. Aqueous Two-Phase Systems," Vol. 228, pp. 403-411, H. Walter and G. Johansson, eds., Academic Press, Inc., San Diego, CA (1994).

35. C. Larsson, Isolation of highly purified intact chloroplasts and of multiorganelle complexes containing chloroplasts, *in*: "Methods in Enzymology. Aqueous Two-Phase Systems," Vol. 228, pp. 419-424, H. Walter and G. Johansson, eds., Academic Press, Inc., San Diego, CA (1994).

36. D.J. Morré, T. Reust, and D.M. Morré, Plasma and internal membranes from cultured mammalian cells, *in*: "Methods in Enzymology. Aqueous Two-Phase Systems," Vol. 228, pp. 448-450, H. Walter and G. Johansson, eds., Academic Press, Inc., San Diego, CA (1994).

37. A. Persson and B. Jergil, Rat liver plasma membranes, *in*: "Methods in Enzymology. Aqueous Two-Phase Systems," Vol. 228, pp. 489-496, H. Walter and G. Johansson, eds., Academic Press, Inc., San Diego, CA (1994).

38. P.-Å. Albertsson, Domain structure of biological membranes obtained by fragmentation and separation analysis, *in*: "Methods in Enzymology. Aqueous Two-Phase Systems," Vol. 228, pp. 503-511, H. Walter and G. Johansson, eds., Academic Press, Inc., San Diego, CA (1994).

39. P.-Å. Albertsson, E. Andreasson, H. Stefánsson, and L. Wollenberger, Fractionation of thylakoid membrane, *in*: "Methods in Enzymology. Aqueous Two-Phase Systems," Vol. 228, pp.469-482, H. Walter and G. Johansson, eds., Academic Press, Inc., San Diego, CA (1994).

40. L. Hammar, Concentration of biomaterials: virus concentration and viral protein isolation, *in*: "Methods in Enzymology. Aqueous Two-Phase Systems," Vol. 228, pp. 640-658, H. Walter and G. Johansson, eds., Academic Press, Inc., San Diego, CA (1994).

41. K.-E.I. Magnusson, Testing for charge and hydrophobicity correlates in cell-cell adhesion, *in*: "Methods in Enzymology. Aqueous Two-Phase Systems," Vol. 228, pp. 326-334, H. Walter and G. Johansson, eds., Academic Press, Inc., San Diego, CA (1994).

HYDROPHOBIC AND CHARGE EFFECTS IN THE PARTITIONING OF SOLUTES IN AQUEOUS TWO-PHASE SYSTEMS

Mark A. Eiteman

Department of Biological and Agricultural Engineering
Driftmier Engineering Center
University of Georgia
Athens, GA 30602

INTRODUCTION

Dobry and Boyer-Kawenoki[1] in 1947 were the first to provide a "methodical study of compatibility in polymer solutions" that had been noted in the study of paints and varnishes. They found that a solution of two polymers dissolved in a single solvent (prepared by mixing two single-polymer solutions) often formed two liquid phases. In their survey, Dobry and Boyer-Kawenoki also noted that each phase in such a system contained principally, but not exclusively, one of the polymers. They eloquently theorized the nature of phase separation, stating, "in a solution of [polymer] A enough free space is available for A molecules, although there is none for [polymer] B molecules."

Albertsson shortly thereafter recognized the potential of analogous water-based systems for the separation of biomolecules.[2] Such "aqueous two-phase systems" or "aqueous biphasic systems" form when an aqueous solution exceeds specific threshold concentrations of two water soluble but mutually incompatible components. Any solute, a biomolecule such as a protein being of particular interest, added to these biphasic systems will distribute between the two aqueous phases. The important engineering parameter describing this phenomenon is a solute's partition or distribution coefficient, denoted K, and it is defined as the solute's upper phase concentration divided by its lower phase concentration. The term *component* is reserved for the compounds dissolved in the solution which participate in the phase-forming phenomenon, while *solute* connotes a dissolved species sufficiently dilute not to influence the phase-forming components. These two academic terms may in practice be difficult to distinguish, as even a small amount of a solute may alter the system itself, particularly if it is a high molecular weight protein.

Many pairs of soluble polymers, including poly(ethylene glycol) (PEG), poly(propylene glycol), dextran, ficoll, methyl cellulose, poly(vinyl alcohol), and various co-polymers, when dissolved together in water form aqueous two-phase systems. The combination of one polymer and a water soluble salt such as the sodium, potassium, ammonium or magnesium salts of sulfate, phosphate or citrate, will often also form a

Aqueous Biphasic Separations: Biomolecules to Metal Ions
Edited by R.D. Rogers and M.A. Eiteman, Plenum Press, New York, 1995

31

biphasic system. When one considers the wide range of polymer molecular weights available, the multitude of possible systems from which one can make a selection seems unmanageable.

Two fundamental types of models exist to aid the selection of an aqueous two-phase system. One type of model is used to predict the phase diagram of the particular mixture. Such a model can be used to predict the binodal, the boundary between the one-phase and the two-phase regions on axes of component concentrations. Such models are naturally founded on thermodynamic principles, and can become quite complex. Nevertheless, models considering fluctuation solutions theory,[3-5] polymer lattice theories,[6] UNIQUAC,[7,8] osmotic virial expansions,[9-11] polymer scaling,[12] statistical geometry[13] and others[14,15] have afforded great insights into phase formation. In general, these models suffer from a lack of appropriate experimental data from the numerous possible systems to determine values for the various model parameters. The models often do not simplify the daunting task of selecting for a particular application a system from among the numerous possibilities.

In contrast to predicting the binodal, the goal of a second type of model is to predict the partition coefficient. Fortunately, the same thermodynamic models used to predict the binodal can often be extended to the prediction of a solute's partition coefficient. Such extensions essentially involve analogous thermodynamic arguments but consider a four-component system (water/component A/component B/solute) instead of a three-component system. While several of these model extensions have been successful at predicting partition coefficients, because of the infinite possible phase system and solute combinations, they are currently impractical for selecting an optimal phase system for a given separation. Moreover, even the ideal but computationally complex case of a four-component system is rarely achieved in practice, since most separations—by definition—involve numerous, perhaps unknown impurities of inconsistent concentration.

With these two types of rigorous models in mind, another approach toward the ultimate goal of selecting a phase system for a particular separation has been merely to accumulate many observations and develop practical heuristics. This practice over the last 35 years has caused aqueous two-phase system selection to appear to be more of an art than a science. Nevertheless, the accumulated observations concerning aqueous two-phase systems constitute a significant knowledge-base, which might be used to develop neural networks[16] or fuzzy systems for selecting appropriate phase systems and conditions. Such approaches can also identify where the knowledge-base is sparse. Specific quantifiable factors already identified which influence a solute's partition coefficient include solute hydrophobicity,[17-19] polymer molecular weight,[20-22], system temperature,[22,23] system pH,[24-30] solute charge[29,30] and the presence of additional salts.[30-36] One example of the value of such an accumulated knowledge-base is several independent observations that partition coefficients of negatively charged solutes decrease in PEG/phosphate or PEG/sulfate systems as the alkali halide concentration is increased. In separate studies, this observation has been made for thaumatin in PEG/phosphate and PEG/sulfate systems,[33] and for ovalbumin[34] and α-amylase[35] in the PEG/phosphate system. Conversely, the partition coefficients of positively charged solutes, such as vancomycin[36] and lysozyme[34] in the PEG/phosphate system, increase with increasing alkali halide concentration. From these observations, one might construct very practical heuristics for affecting the partition coefficient of a given solute. An understanding of why this observed phenomenon occurs would further improve the heuristics and lead to a quantitative model of how much halide is needed to afford a desired change in the partition coefficient. Although such "models" are not derived from purely theoretical considerations, they should nevertheless be thermodynamically consistent.

In our research program we have begun with observations and heuristics of others to formulate simple models. Our evolving knowledge-base on partitioning phenomena has led to improved quantification using models which aid the *a priori* selection of an aqueous two-phase system. Attention is focussed on the importance of solute hydrophobicity and solute charge, particularly for low molecular weight solutes, and how these phenomena relate to phase system parameters to influence partitioning. Although these evolving models do not account for all complex partitioning behavior, they often do lead to practical selection of phase systems and operating parameters.

An appropriate starting place is a simple equation used by Diamond and Hsu.[37] This equation may also be derived from the osmotic pressure virial expansion[9] by recognizing that the tie-lines of binodals for the vast majority of aqueous two-phase systems tend to be parallel:

$$RTlnK = k\Delta w_2 \qquad (1)$$

This equation states that the logarithm of a solute's partition coefficient is proportional to the concentration difference of component 2 (often PEG) between the phases, or the tie-line length. Equation 1 leads to a heuristic that doubling the tie-line length of the system will square the partition coefficient of any solute, *if increasing the tie-line length does not alter the proportionality constant, k.* The goals of our research become to understand the proportionality constants of numerous solutes in phase systems. Specifically, how do a solute's hydrophobicity and charge, and the selection of a particular phase system, influence the proportionality constant in Equation 1 to affect partitioning?

Hydrophobicity

Few terms are so broadly applied while being so informally understood as "hydrophobicity." Since hydrophobicity cannot strictly be measured like pH, the term hydrophobicity is often descriptive rather than quantified. Phenomena associated with hydrophobicity are instead correlated with a measurable phenomenon. In the study of pharmaceuticals, for example, hydrophobicity is often correlated with a solute's partitioning between water and an arbitrary, non-aqueous phase such as octanol. Solute hydrophobicity has also been correlated with elution using hydrophobic interaction chromatography. Although numerous molecular interactions contribute to octanol partitioning or chromatographic results besides hydrophobicity, both correlations have been very useful.

The goal of this chapter is not to elaborate on a formal definition for hydrophobicity. For a hopefully sufficient understanding of the term for the current topic, one might recognize that for any homologous series of uncharged solutes such as normal alcohols, the solutes become less "water-like" as the number of methylene groups is increased. Thus, pentanol is more hydrophobic than butanol, which is itself more hydrophobic than propanol. This increased hydrophobicity with alcohol size is manifested in observations such as decreased aqueous solubility and increased partitioning into octanol from water. Although these phenomena may be related to a Gibbs free energy of transfer of a methylene group, the relationship between pentanol and butanol is probably different from the relationship between methyl tryptophan and tryptophan, both of which vary in one methylene group. Moreover, the relationship between butanol and tryptophan is probably different in octanol-water than in hexane-water. Nevertheless, hydrophobicity scales for solutes may be and have been constructed by extension of such arguments.[38]

Like in octanol-water systems, uncharged solutes of differing hydrophobicity partition differently in a given aqueous two-phase system. In general, the PEG phase (usually the upper phase) is the more hydrophobic of the two phases. Thus, butanol has a greater partition coefficient than propanol in these aqueous two-phase systems. In these majority of cases, the proportionality constant in Equation 1 therefore increases with increasing solute hydrophobicity. However, the partition coefficients of an uncharged solute in two different two-phase systems of identical Δw_2 in general are unequal. That is, butanol in general will have a different partition coefficient in system A than it has in a system B having the same tie-line length, Δw_2. This observation implies that some intrinsic hydrophobicity difference exists between any two aqueous phases, and that the intrinsic nature of a phase system itself affects the value of the proportionality constant in Equation 1. Not only does the proportionality constant depend on the particular solute, but it depends on the phase system. This notion is supported by the extended osmotic virial equation[9] which shows that the coefficients of each concentration difference term are interaction parameters between the solute and phase-forming components.

For a given two-phase system, we can find an uncharged solute which partitions equally between the phases for all tie-line lengths. The hydrophobicity of this solute provides us with a convenient means to quantify the hydrophobicity of the phases themselves. Thus, a system in which butanol partitions equally between the phases may be thought of as intrinsically more hydrophobic than a phase system for which propanol has a partition coefficient of one. Such arguments lead to an equation of the form:[39]

$$RTlnK = j(\alpha + \Delta f)\Delta w_2 \qquad (2)$$

Here, Δf is the solute's hydrophobicity, a value which may be related to the Gibbs free energy of transfer of the solute between the phases. The parameter j is a type of discrimination factor, indicating the difference in partition coefficients that is obtained by a unit change in solute hydrophobicity. The parameter α is related to that instrinsic hydrophobicity of the phase system, and will be merely referred to as the phase constant. In a phase system for which an uncharged solute's hydrophobicity equals $-\alpha$, the partition coefficient will be one for all values of Δw_2. To obtain a different partition coefficient for such a solute, one must either select a different system with a different α, change the solute's charge if present (to be discussed), or rely on specific interactions (e.g., ligand partitioning).

Equation 2 permits phase systems to be classified by the values of the two parameters, j and α, both of which can readily be obtained by a series of experiments (such as partitioning normal alcohols). Moreover, the equation has the advantage of being written so that the solute hydrophobicity may be considered to be additive.[18] This consequence permits one to predict the partition coefficient of larger solutes (e.g., peptides) from partition coefficients of constituent solutes (e.g., amino acids).[39]

Charge

In order to study the effect of charge on the partitioning of solutes, it should be isolated from other confounding effects such as hydrophobicity. The ideal method to study charge alone would be to compare the partitioning of two compounds, one charged and the other uncharged, but otherwise *identical*, in an aqueous two-phase system. These experiments could lead to some understanding of how the proportionality constant in Equation 1 is affected by a solute's charge, as well as how this phenomenon is manifested in different aqueous two-phase systems (i.e., does a phase system have an instrinsic "charge"?). The idealized experiment of partitioning two, but for charge, identical compounds unfortunately cannot be performed. This experiment may merely be approached by partitioning similar compounds in a single system such as nitrobenzyl alcohol and nitrobenzyl acid.[28]

Another aspect of the charge effect as it relates to partitioning in aqueous two-phase systems is that the charge of any solute depends on the pH of the solution. The charge of any compound is expressed in terms of an equilibrium between the various charged species. Thus, a simple organic acid such as acetic acid is uncharged at low pH, a pH at which it might behave no differently than a hypothetical uncharged solute of identical hydrophobicity. However, as the pH increases, so does the proportion of acetic acid existing as an ion (acetate). Thus, as the pH increases, the two species of acetic acid partition, and one must predict the partition coefficient of each species to determine an overall partition coefficient. The envisioned computations become necessarily complex for a solute of multiple dissociations. If charge does indeed affect partitioning in aqueous two-phase systems, the partition coefficient of acetic acid should deviate more and more as pH increases from the partition coefficient of the hypothetical uncharged solute of identical hydrophobicity. Such observations have been made for a few solutes.[28-30] In practice, the proposed experiment of changing the pH of a phase system solely to study charge effects has the unfortunate consequence of shifting the phase diagram and hence altering the values of j, α, and Δw_2 in Equation 1.

Recently, a pH difference between the phases of an aqueous two-phase system was shown to be a phase system parameter which influences the partitioning of charged solutes.[28-30] Using mass balances for all species in a phase system, along with the definition of the partition coefficient, equations were derived to predict the partition coefficient of a charged solute relative to the partition coefficient of the neutral species alone. The resulting model indicates that the partition coefficient of a charged solute relative to its neutral species depends on the dissociation constants of the solute and the pH difference between the phases.[30] The essential heuristic contained in the model is that the greater the pH difference between the phases, the greater the difference between the partition coefficients of otherwise identical charged and uncharged solutes. Specifically, a negatively charged solute will partition more into the phase having greater pH than a neutral solute, which will in turn partition more into this more basic phase than a positively charged solute. In other words, if the pH difference between the phases is positive (upper phase more basic), a negatively charged solute will have a greater partition coefficient than a neutral solute, which will have a greater partition coefficient than a positively charged solute (each of identical hydrophobicity). If the pH difference of such a phase system (with positive ΔpH) is reduced without affecting the tie-line length or the intrinsic hydrophobicity, then the partition coefficient of a negatively charged solute is predicted to decrease, and a positively charged solute is predicted to increase. Naturally, the partition coefficient of a neutral solute remains the same, as its partitioning is not influenced by a pH difference between the phases.

These hydrophobicity and charge effects suggest that quite complicated partitioning behavior is possible in aqueous two-phase systems. For example, in a system having a pH of about 6 with a large positive pH difference (perhaps 0.5 units), a negatively charged solute could have a partition coefficient identical to a neutral solute of greater hydrophobicity. If the pH of this system is increased to 8, and in the process the pH difference decreases and the tie-line length increases, then the partition coefficient of the neutral solute will increase (due to the greater tie-line length). However, the resulting partition coefficient of the negatively charged solute will depend upon the opposing influence of hydrophobicity (which would tend to increase the partition coefficient) and charge (which would tend to decrease the partition coefficient). In contrast, if the pH of the original system is decreased to 4, and in this case the pH difference increases while the tie-line length decreases, then the partition coefficient of the neutral solute will decrease (since tie-line length is decreased). Because of its own dissociation, however, the originally negatively charged solute may become neutral or even positively charged as pH is decreased. If this solute remains negatively charged, then the resulting partition coefficient will depend upon the opposing influence of hydrophobicity (which would decrease the solute's partition coefficient) and charge (which would increase the solute's partition coefficient). However, if this solute becomes neutral or positively charged, then the partition coefficient would decrease from both influences. In fact, the greatest partition coefficients occur when hydrophobic and charge effects act in concert.

The objectives of this chapter are to illustrate the hydrophobic and charge effects and their influences in example aqueous two-phase systems. Three series of PEG/citrate systems are used to illustrate these effects. One series has a pH of about 5.8, the second a pH of about 8.5, and the third a pH of 8.3, identical to the second but with sodium chloride substituted for a portion of the sodium citrate.

MATERIALS AND METHODS

Three different ratios of sodium citrate dihydrate, citric acid and sodium chloride (Sigma Chem. Co., St. Louis, USA) were used for the preparation of the two-phase systems used in this study: 8.5% sodium citrate/0.5% citric acid/0.0% sodium chloride, 9.0%/0.0%/0.0%, and 7.0%/0.0%/2.0% (percentages indicate weight fractions). For each of these ratios, four 10 mL systems were prepared by varying the percentage of PEG-8000 (Sigma Chem. Co., St. Louis, USA) from 15.0% to 19.5%.

All sets of 12 phase systems studied were agitated frequently for two days, equilibrated at 25.0°C for five days, then carefully separated with Pasteur pipets immediately before analysis. The pH of each phase was measured as reported previously,[28] and the PEG concentration was measured by the method of Skoog.[40]

The partitioning of several solutes was studied in these aqueous two-phase systems including alcohols, amino acids, peptides and proteins. In each case, approximately 5 mg of a single solute were added to 10 ml of each two-phase system studied. These phases were similarly agitated and equilibrated at 25.0°C, before being separated and analyzed. Butanol, pentanol and hexanol (Aldrich Chemical Co., Milwaukee, USA) were analyzed with gas chromatography. Alanine and glycine (Sigma Chemical Co., St. Louis, USA) were analyzed by their reaction with ninhydrin and absorbance at 570 nm. Tryptophan, tryptophan-glycine, tryptophan-glycine-glycine, tryptophan-phenylalanine and tryptamine (Sigma Chemical Co.) were analyzed by their absorbance at 278 nm. Phenylalanine, leucine-phenylalanine and tyrosine-phenylalanine (Sigma Chemical Co.) were each analyzed by their absorbance at 256 nm. Lysozyme and papain (Sigma Chemical Co.) were analyzed by the Lowry protein assay (Sigma Chemical Co.).

RESULTS AND DISCUSSION

The first step in understanding any aqueous two-phase system is to understand the concentrations at which an aqueous two-phase system occurs for the particular temperature and components. In the case of the PEG/sodium citrate/citric acid/sodium chloride phase system at 25°C, Table 1 shows the concentration differences observed for the different ratios of components used. The results are what is commonly observed in other PEG/salt systems. In comparing the NaCl-free phase system at low pH (i.e., the 8.5/0.5/0.0 system) with the more basic NaCl-free system (i.e., the 9.0/0.0/0.0 system), the tie-line length is greater for the more basic system. In other words, the binodal extends closer to the zero concentration axes the higher the pH. This is true when the concentration axes are written on a weight basis or on a mole basis. Of course, considering Equation 1 these observations suggest that the partition coefficient of any solute will change with changing pH, merely because changing pH alters the phase diagram. For the partition coefficient of an uncharged solute to remain the same as pH is changed, the concentration of the phase forming components would have to be adjusted to maintain the same tie-line length. This adjustment would itself also need refinement due to potential changes in the discrimination factor and phase constant (Equation 2) with pH.

When sodium chloride is substituted for sodium citrate on a weight basis in the

Table 1. Concentration difference ($\Delta w_2 = w_2' - w_2''$) between the phases in three aqueous two-phase systems. The composition of each component is given as its weight fraction.

PEG-8000 (weight%)	Sodium Citrate / Citric Acid / NaCl (weight%)		
	8.5 / 0.5 / 0.0	9.0 / 0.0 / 0.0	7.0 / 0.0 / 2.0
15.0	0.142	0.215	0.192
16.5	0.195	0.257	0.247
18.0	0.213	0.275	0.261
19.5	0.258	0.317	0.306

preparation of the two-phase system, the tie-line length decreases as shown in Table 1. Sodium chloride is less effective than sodium citrate at causing the two-phase phenomenon. If instead 2.0% is added to any of the 9.0% sodium citrate systems (data not shown), the tie-line length increases slightly. Like adjusting the pH, merely adding some additional chloride or other salt will shift the phase diagram and hence affect the partitioning of a neutral solute. Tie-line lengths, by whatever consistent measurement used, will enable the calculation of a phase constant and discrimination factor by partitioning studies.

As discussed in the introduction, the partitioning of charged solutes is affected by the pH of each of the phases. Table 2 shows the measured pH difference for each of the 12 phase systems. The data show that the pH difference measured in the 8.5/0.5/0.0 and the 9.0/0.0/0.0 systems increases with increasing tie-line length. Interestingly, the pH difference observed in this experiment appears to be independent of the pH of the phase system itself. That is, the increase in pH difference can be correlated to the increase in tie-line length, and is not observed to be a function of strictly pH.

For the 7.0/0.0/2.0 system, the pH difference between the phases is much lower than for the system without sodium chloride. In this chloride system, ΔpH is not observed to be a function of tie-line length (although the estimated error in ΔpH of 0.05 units prevents great assurance). The decrease in pH difference when NaCl is added cannot be accounted for merely by the change in the tie-line length. This decrease in ΔpH is a phenomenon

Table 2. pH difference (ΔpH = pH' - pH'') between the phases in three aqueous two-phase systems. The composition of each component is given as its weight fraction.

PEG-8000 (weight%)	Sodium Citrate / Citric Acid / NaCl (weight%)		
	8.5 / 0.5 / 0.0	9.0 / 0.0 / 0.0	7.0 / 0.0 / 2.0
15.0	0.04	0.14	0.04
16.5	0.16	0.16	0.02
18.0	0.17	0.24	0.07
19.5	0.24	0.28	0.04

particular to the addition of salts such as sodium chloride. Previous results,[30] in fact, have shown that ΔpH decreases continuously with increasing sodium chloride for the PEG/potassium phosphate system, reaching essentially zero when the sodium chloride concentration reaches 2.0M. Unpublished results obtained here have shown an even more marked decrease in ΔpH when sodium iodide is added to the phosphate phase system.

The next step in understanding these three different sets of two-phase systems is to quantify the hydrophobicity of each phase system. The details of this quantification will not be discussed as they are thoroughly covered elsewhere.[39] Briefly, the technique involves rewriting Equation 2 for normal alcohols:

$$RT \ln K = j\,(\alpha + \Delta f_{MeOH} + N\Delta f_{CH_2})\,\Delta w_2 \qquad (3)$$

In Equation 3, the hydrophobicity of a normal alcohol is merely $\Delta f_{MeOH} + N\Delta f_{CH2}$, where N is the number of methylene (CH_2) groups on the alcohol beyond methanol. Δf_{CH2} is the hydrophobicity of a methylene group (a previously determined value of 500 cal/mol[41] will be used). Equation 3 permits the calculation of the discrimination factor, j, from partitioning a series of normal alcohols—the parameter is directly related to the slope of $\ln K$ versus N.

Figures 1-3 show the partition coefficients of butanol, pentanol and hexanol in each of the three sets of PEG/sodium citrate/citric acid/sodium chloride phase systems. Fitting Equation 3 to these data results in values of j of 1.74 for the 8.5/0.5/0.0 system ($r^2 = 0.970$, 12 data points), 1.40 for the 9.0/0.0/0.0 system ($r^2 = 0.958$, 12 data points) and 1.40 for the 7.0/0.0/2.0 system ($r^2 = 0.953$, 12 data points). There is roughly a 20% decrease in the "discrimination power" of the aqueous two-phase system in increasing the pH of this phase system from 5.8 to 8.5. This does not mean the more basic phase system cannot accomplish a separation as well, it merely means that in the more basic system a greater tie-line length is necessary to discriminate between a particular pair of neutral solutes. The discrimination factor does not change by replacing some of the sodium citrate with sodium chloride. This observation means that two systems, one with and a second without sodium chloride, having identical tie-line length will result in the same discrimination between a particular pair of solutes. Note that in order to have two such systems of identical tie-line length, the system with sodium chloride will have to have the greater concentration of PEG. This observation does not, by itself, mean that two solutes will partition identically in the two systems, since the phase constant also affects partitioning.

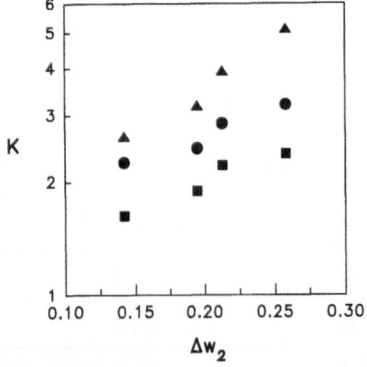

Figure 1. Partition coefficients (K) of butanol (■), pentanol (●) and hexanol (▲) in the 8.5/0.5/0.0 sodium citrate/citric acid/NaCl two-phase system.

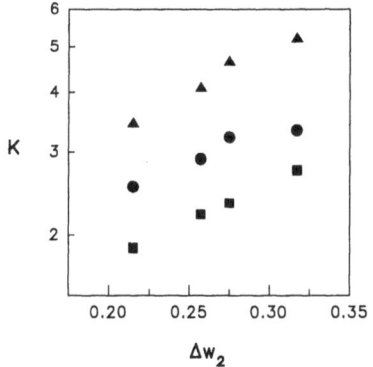

Figure 2. Partition coefficients (K) of butanol (■), pentanol (●) and hexanol (▲) in the 9.0/0.0/0.0 sodium citrate/citric acid/NaCl two-phase system.

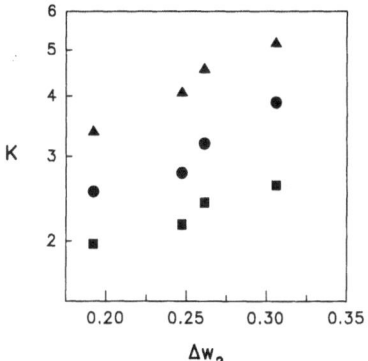

Figure 3. Partition coefficients (K) of butanol (■), pentanol (●) and hexanol (▲) in the 7.0/0.0/2.0 sodium citrate/citric acid/NaCl two-phase system.

In order to determine the second parameter in Equation 2, α, glycine must be partitioned in a phase system in which this solute is neutral. Only near this solute's isoelectric point can the phase constant be measured without concern that charge is affecting the observed partitioning data of glycine. Figure 4 shows the results of partitioning glycine and other amino acids in the 8.5/0.5/0.0 system, which has a pH near the isoelectric points of these amino acids. Since the hydrophobicity of glycine has been assigned a value of zero, Equation 2 for this solute merely simplifies to:

$$RT\ln K = j\alpha\Delta w_2 \qquad (4)$$

and the value of α may be determined from the slope of the glycine partitioning data. From these data in Figure 4, α was calculated to be -1450 cal/mol ($r^2 = 0.931$, 4 data points). At this point, by reconsidering the best-fit of Equation 3 to the data in Figure 1, the value of Δf_{MeOH} (the only remaining unknown in this equation) can be calculated to be 1090 cal/mol. Since this 8.5/0.5/0.0 system has a phase constant of -1450 cal/mol,

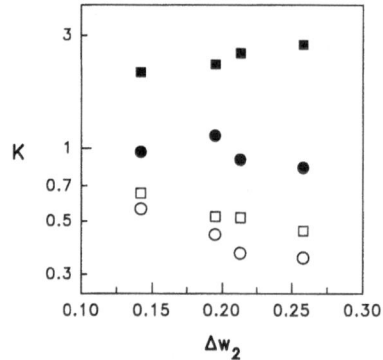

Figure 4. Partition coefficients (K) of trytophan (■), phenylalanine (●), alanine (□) and glycine (○) in the 8.5/0.5/0.0 sodium citrate/citric acid/NaCl system.

Equation 2 predicts the methanol will partition in this phase system according to the equation

$$\ln K = -1.06 \Delta w_2 \qquad (5)$$

Also with this value of Δf_{MeOH}, the value of the phase constant can be <u>estimated</u> for the other two systems from the partition data for the normal alcohols shown in Figures 2 and 3. (Determining the phase constant cannot be accomplished by partitioning glycine because it is charged at the pH of these other two systems.) For the 9.0/0.0/0.0 system, the value of α was estimated to be -1330 cal/mol, while for the 7.0/0.0/2.0 system the value of the phase constant was estimated to be -1230 cal/mol.

From all these calculations a few comments can be made to clarify what we have learned about the three phase systems with regard to the influence of solute hydrophobicity on partitioning. By way of illustration, consider two neutral solutes we wish to separate from each other, one (A) having a hydrophobicity of $\Delta f = 1000$ cal/mol, and the second (B) a hydrophobicity of $\Delta f = 2000$ cal/mol. According to our results, the predicted partition coefficients of A and B will follow the equations listed in Table 3. The last column in

Table 3. Equations to predict the partitioning of two hypothetical solutes in three PEG/sodium citrate/citric acid/sodium chloride aqueous two-phase systems. Solute "A" has a relative hydrophobicity of 1000 cal/mol, while solute "B" has a relative hydrophobicity of 2000 cal/mol.

System	Solute "A"	Solute "B"	$\beta_{B,A}$
8.5/0.5/0.0	ln K = -1.32 Δw_2	ln K = 1.62 Δw_2	exp(2.94Δw_2)
9.0/0.0/0.0	ln K = -0.78 Δw_2	ln K = 1.58 Δw_2	exp(2.36Δw_2)
7.0/0.0/2.0	ln K = -0.54 Δw_2	ln K = 1.82 Δw_2	exp(2.36Δw_2)

Table 3 shows the selectivity, $\beta_{B,A} = K_B/K_A$, as a function of tie-line length for the separation between solute B and solute A.

As Table 3 shows, in all three systems solute A is predicted to have partition coefficients less than one, while solute B is predicted to have partition coefficients greater than one. This result arises because solute A has a hydrophobicity lower than the negative of the phase constant, while solute B has a hydrophobicity greater than -α for all three systems. For a given tie-line length, solute A will have the lowest partition coefficient in the 8.5/0.5/0.0 system, a higher partition coefficient in the 9.0/0.0/0.0, and the greatest (but still less than one) partition coefficient in the 7.0/0.0/2.0 system. In contrast, solute B will have the lowest partition coefficient in the 9.0/0.0/0.0 system, slightly higher in the 8.5/0.5/0.0 system, and the greatest partition coefficient in the 7.0/0.0/2.0 system. A more important parameter in separations than the partition coefficient is the selectivity, which is the ratio of the partition coefficients of two solutes to be separated. As the last column in Table 3 shows, these two solutes are separated best at a given tie-line length in the 8.5/0.5/0.0 system. In this example the 9.0/0.0/0.0 and the 7.0/0.0/2.0 systems have identical selectivity in the separation of these two solutes. In order to achieve the same separation as the 8.5/0.5/0.0 system in one of these two system, the tie-line length would have to be increased by 25%. One must also recall that the identical concentration of PEG in the three systems does not yield identical tie-line lengths (see Table 1). After considering cost of PEG, potential difficulties in handling highly viscous phases, and mass transfer differences, it may indeed turn out that the best selection is a phase system of lower selectivity. Nevertheless, these relationships have begun to quantify partitioning and will aid in the selection of the phase systems. All these equation should be considered empirical, but they do give a good start to selecting an optimal system for a given separation.

Figure 4 also shows the partitioning of three amino acids other than glycine. Considering the structures of the amino acids, one should not be surprized that tryptophan has the greatest partition coefficients, then phenylalanine, and alanine slightly greater than glycine. The observed hydrophobicities of these three amino acids can be calculated from their partitioning data. From these calculations, tryptophan is found to have a hydrophobicity of 2870 cal/mol, phenylalanine 1360 cal/mol, and alanine 390 cal/mol. The value previously assumed for the hydrophobicity of a methylene group, 500 cal/mol, may be compared to the hydrophobicity of alanine, which theoretically should be identical since alanine differs from glycine by one methylene group.

The next complication one might make towards the goal of understanding partitioning phenomena is the partitioning of peptides. Figure 5 shows the partition coefficients of five small peptides in the 8.5/0.5/0.0 system. These peptides are close to neutral at the pH of this system. Calculating the observed hydrophobicity for these five peptides by Equation 2, we find that trp-phe is the most hydrophobic with a Δf of 4060 cal/mol, then are tyr-phe at 3170 cal/mol, trp-gly at 2540 cal/mol, trp-gly-gly at 2300 cal/mol and leu-phe at 1070 cal/mol. The first four peptides might be referred to as relatively hydrophobic since they partition preferentially into the upper phase, while leu-phe might be called relatively hydrophilic since it partitions into the lower phase. The hydrophobicity order of these peptides corresponds with what one might intuit in knowing the structure of the residues.

One might be tempted to use the additivity rule for predicting hydrophobicity of peptides from the hydrophobicities of amino acids. Thus, the data and results from Figure 5 lead to a prediction that trp-phe has a hydrophobicity of 4230 cal/mol based on adding the hydrophobicity of tryptophan and phenylalanine (2870 + 1360). For trp-gly and trp-gly-gly this prediction is 2870 cal/mol, since glycine has been assigned a hydrophobicity of zero. The difference between the measured and predicted hydrophobicities is 170 cal/mol

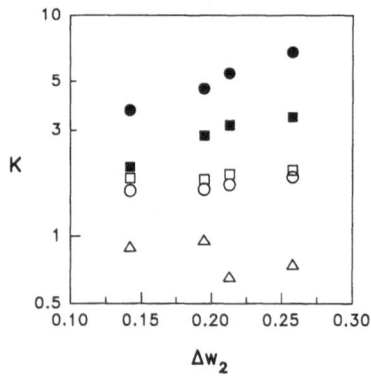

Figure 5. Partition coefficients (K) of trp-phe (●), tyr-phe (■), trp-gly (□), trp-gly-gly (○) and leu-phe (△) in the 8.5/0.5/0.0 sodium citrate/citric acid/NaCl system.

for trp-phe, 420 cal/mol for trp-gly and 285 cal/mol for each peptide bond formed for trp-gly-gly. In all cases the predicted hydrophobicity is greater than the measured hydrophobicity, an observation made previously for numerous other peptides. If this difference is a real phenomenon, it can be interpreted in two obvious ways. The difference may be due to a loss of hydrophobicity associated with interactions between adjacent residues on a peptide. This interaction would be expected to increase with increasing molecular size from an amino acid to peptide, and could become quite significant for larger peptides. The difference might also be due to the water molecule released during each peptide bond formation. Strictly speaking, trp-phe does not have the same atoms as a tryptophan and phenylalanine.

Up to this point the results and discussion have been limited to the effect that hydrophobicity has on partitioning. Each of the peptides shown in Figure 5, however, can have a charge in phase systems of other pH. Preferably, one could *a priori* select the system of ideal pH. With knowledge of the pH and pH difference (Table 2) of these three systems, and using the heuristics clarified in the introduction, the trend in partition coefficients of each of these peptides can be predicted in the other two phase systems.

For the 9.0/0.0/0.0 system, the pH is 8.5 and the pH difference between the phases is always greater than 0.10 units. Each of the peptides is negatively charged at this pH. As noted in the introduction, negatively charged solutes are expected to partition more than neutral solutes into the phase having greater pH. Thus, each of these peptides is predicted to have a higher partition coefficient in the 9.0/0.0/0.0 system than it would have *had it remained neutral*. We must first consider how these solutes would have partitioned had they remained neutral. From hydrophobicity considerations alone as discussed previously, relatively hydrophobic solutes (e.g., solute B in Table 3) are predicted to have slightly lower partition coefficients in the 9.0/0.0/0.0 system that in the 8.5/0.5/0.0 system at a given tie-line length. However, for a given concentration of PEG, the 9.0/0.0/0.0 system yields a greater tie-line length than the 8.5/0.5/0.0 system (Table 1). Considering first hydrophobicity *alone*, at a given PEG concentration the five peptides in Figure 5 are predicted to have only slightly higher partition coefficients in the 9.0/0.0/0.0 system than in the 8.5/0.5/0.0 system. Since charge effects serve to increase the partition coefficients of negatively charged solutes when ΔpH is positive, these five peptides are predicted to have significantly greater partition coefficients in the 9.0/0.0/0.0 system. The change in partitioning with increased pH is predicted to be almost entirely controlled by the charge effect.

Figure 6 shows the partition coefficients of the five peptides in the 9.0/0.0/0.0 system. Each of these peptides shows a 100% or greater increase in the partition coefficients. The partition coefficient of the most hydrophilic peptide studied, leu-phe, has increased to slightly above one. The partition coefficient of the most hydrophobic, trp-phe, has increased from 6.8 to 50 in the system of greatest tie-line length. These solutes have not become "more hydrophobic" as might be inferred by their increased partition coefficient. Their negative charges have caused shifts in the resulting partition coefficients.

For the 7.0/0.0/2.0 system, the pH is about 8.3 but the pH difference between the phases is less than in the 9.0/0.0/0.0 system. Each of the peptides is still negatively charged, but the charge effect is predicted to be less important in partitioning than in the 9.0/0.0/0.0 system. Once again, we must first consider hydrophobic effects, however. For a given tie-line length, the logarithms of the partition coefficients are predicted to be about 15% greater in the 7.0/0.0/2.0 system than in the 9.0/0.0/0.0 system (see Table 3). However, for a given PEG concentration, the tie-line length is about 5-10% less in the 7.0/0.0/2.0 system than in the 9.0/0.0/0.0 system. Thus, based on hydrophobicity considerations the partition coefficients of these peptides are predicted to be slightly greater in the 7.0/0.0/2.0 system than in the 9.0/0.0/0.0 system. However, since the charge effect is significantly less important in the 7.0/0.0/2.0 system, the partition coefficients of the five peptides are predicted to be less in the 7.0/0.0/2.0 system than in the 9.0/0.0/0.0 system.

Figure 7 shows the partition coefficients of the five peptides in the 7.0/0.0/2.0 system. When compared to the 9.0/0.0/0.0 system, each of these peptides shows a 10-50% decrease in the partition coefficients. The partition coefficient of the most hydrophilic peptide studied, leu-phe, is still slightly above one. The partition coefficient of the most hydrophobic, trp-phe, has decreased from 50 to 24.5 in the system of greatest tie-line length. The charge effect, not entirely removed in the 7.0/0.0/2.0 system, has less influence than in the 9.0/0.0/0.0 system.

Three more examples will be considered. The solutes shown in Figures 6-7 were each negatively charged. One might correctly expect contrasting behavior with positively charged solutes. A relatively hydrophobic positively charged solute will have a lowered partition coefficient than an identical neutral solute in any system having a positive pH difference between the phases. Such a solute would also be predicted to show an increase

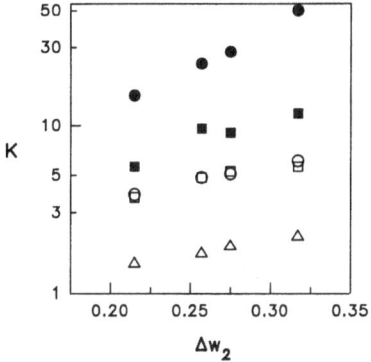

Figure 6. Partition coefficients (K) of trp-phe (●), tyr-phe (■), trp-gly (□), trp-gly-gly (O) and leu-phe (Δ) in the 9.0/0.0/0.0 sodium citrate/citric acid/NaCl system.

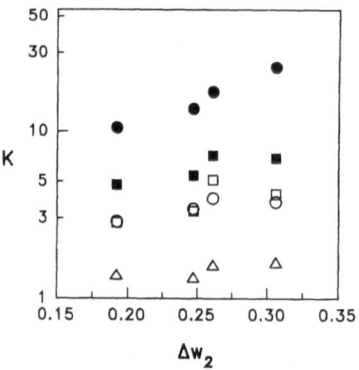

Figure 7. Partition coefficients (K) of trp-phe (●), tyr-phe (■), trp-gly (□), trp-gly-gly (O) and leu-phe (△) in the 7.0/0.0/2.0 sodium citrate/citric acid/NaCl system.

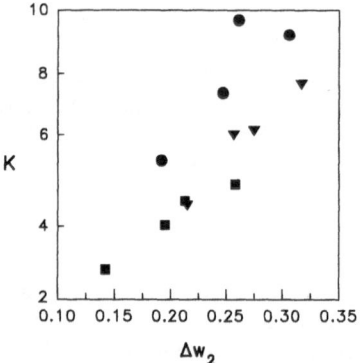

Figure 8. Partition coefficients (K) of tryptamine in the 8.5/0.5/0.0 (■), 9.0/0.0/0.0 (▼), and 7.0/0.0/2.0 (●) sodium citrate/citric acid/NaCl system.

in its partition coefficient as this pH difference between the phases is reduced. Thus, such a solute would have a greater partition coefficient in the 7.0/0.0/2.0 system than in the 9.0/0.0/0.0 system or the 8.5/0.5/0.0 system. Figure 8 shows the partitioning of tryptamine in the series of three systems versus tie-line length. The relationship between the partition coefficient of this solute and the tie-line length is the same for the 8.5/0.5/0.0 and 9.0/0.0/0.0 systems. However, the partition coefficient of tryptamine is significantly greater in the 7.0/0.0/2.0 system, the opposite results of those observed for the negatively charged solutes (Figures 6-7).

Proteins are significantly more complicated than peptides. Additional phenomena not present in the partitioning of small molecules no doubt strongly influence the partitioning of proteins, such as excluded volume effects, refolding, or special interactions between PEG or salt and proteins. Nevertheless, the same charge and hydrophobic effects can serve to aid the selection of an appropriate aqueous two-phase system, particularly for those situations in which hydrophobic and charge effects predominate.

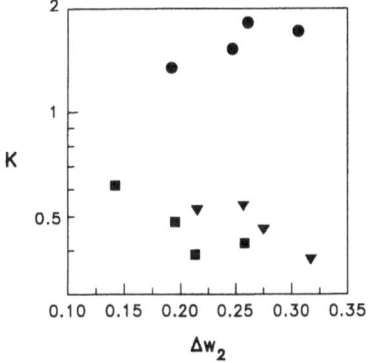

Figure 9. Partition coefficients (K) of lysozyme in the 8.5/0.5/0.0 (■), 9.0/0.0/0.0 (▾), and 7.0/0.0/2.0 (●) sodium citrate/citric acid/NaCl system.

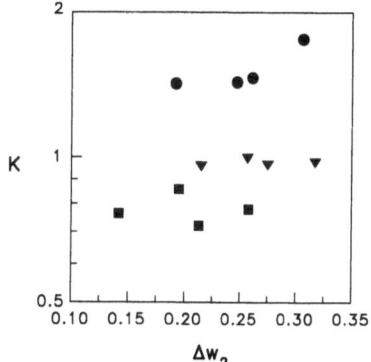

Figure 10. Partition coefficients (K) of papain in the 8.5/0.5/0.0 (■), 9.0/0.0/0.0 (▾), and 7.0/0.0/2.0 (●) sodium citrate/citric acid/NaCl system.

Lysozyme has a molecular weight of 14,400 and an isoelectric point of 10.7. In all the phase systems studied, it would be positively charged. Depending on its hydrophobicity, lysozyme should behave similar to tryptamine in the three series of aqueous two-phase systems studied. Figure 9 shows the results of partitioning lysozyme. In both the 8.5/0.5/0.0 and 9.0/0.0/0.0 systems, lysozyme partitions preferentially to the lower phase. The addition of NaCl significantly increases the partition coefficient of this protein, driving its partition coefficient above one.

Papain is recognized as a relatively hydrophobic protein having a molecular weight of 21,000 and an isoelectric point of 8.75. Figure 10 shows the results of partitioning this protein in the three sets of two-phase systems. Like lysozyme, papain partitions below one in the two sodium chloride-free phase systems, but partitions above one when NaCl is added.

45

CONCLUSIONS

The methodology and heuristics described should provide the basis to begin to select appropriate phase systems for a desired separation. Nevertheless, other potentially important factors are not addressed such as solute-solute interactions, excluded volume effects, and the effects that multiple solutes themselves have on phase system characteristics (e.g., the pH difference).

The most confusing aspect of partitioning phenomena in aqueous two-phase systems is that any change in the system affects both the hydrophobic and charge, and no doubt other, effects. Depending on the magnitude of these two effects, the ultimate partition coefficient may increase, decrease, or remain essentially the same upon a change. To minimize this confusion, each effect can be considered independently.

In the case of systems containing sodium citrate and citric acid, like other PEG/salt systems, increasing the pH by changing the proportion of these two components while maintaining constant PEG concentration increases both the tie-line length and increases the pH difference between the phases.

Acknowledgments

The author acknowledges Joby Miller for experimental support and the Georgia Experiment Stations for financial support.

REFERENCES

1. A. Dobry, and F. Boyer-Kawenoki, Phase separation in polymer solutions, *J. Polymer Sci.*, 2:90 (1947).
2. P.-Å. Albertsson, Partition of proteins in liquid polymer-polymer two-phase systems, *Nature*, 182:709 (1958).
3. H. Cabezas, Jr., J. D. Evans, and D. C. Szlag, A statistical mechanical model of aqueous two-phase systems, *Fluid Phase Equil.*, 53:453 (1989).
4. H. Cabezes, Jr., M. Kabiri-Badr, and D. C. Szlag, Statistical thermodynamics of phase separation and ion partitioning in aqueous two-phase systems, *Bioseparation*, 1:227-233 (1990).
5. H. Cabezas, Jr., J. D. Evans, and D. C. Szlag, Statistical thermodynamics of aqueous two-phase systems, *in:* "Downstream Processing and Bioseparation," ACS Symposium Series 419, J.-F. P. Hamel, J. B. Hunter, and S. K. Sidkar, eds., American Chemical Society, Washington, DC (1990).
6. J. N. Baskir, T. A. Hatton, and U. W. Suter, Thermodynamics of the separation of biomaterials in two-phase aqueous polymer systems: effects of the phase-forming polymers, *Macromolecules*, 20:1300 (1987).
7. C. H. Kang, and S. I. Sandler, Phase behavior of aqueous two-polymer systems, *Fluid Phase Equil.*, 38:245 (1987).
8. C. H. Kang, and S. I. Sandler, A thermodynamic model for two-phase aqueous polymer systems, *Biotechnol. Bioeng.*, 32:1158 (1988).
9. R. S. King, H. W. Blanch, and J. M. Prausnitz, Molecular thermodynamics of aqueous two-phase systems for bioseparations, *AIChE J.*, 34:1585 (1988).
10. C. A. Haynes, H. W. Blanch, and J. M. Prausnitz, Separation of protein mixtures by extraction: thermodynamic properties of aqueous two-phase polymer systems containing salts and proteins, *Fluid Phase Equil.*, 53:463 (1991).
11. D. Forciniti, and C. K. Hall, Theoretical treatment of aqueous two-phase extraction by using virial expansions, *in:* "Downstream Processing and Bioseparation," ACS Symposium Series 419, J.-F. P. Hamel, J. B. Hunter, and S. K. Sidkar, eds., American Chemical Society, Washington, DC (1990).
12. N. L. Abbott, D. Blankschtein, and T. A. Hatton, Protein partitioning in two-phase aqueous polymer systems. 1. Novel physical pictures and scaling-thermodynamic formulation, *Macromolecules*, 24:4334 (1991).

13. Y. Guan, T. E. Treffry, and T. H. Lilley, Application of a statistical geometrical theory to aqueous two-phase systems, *J. Chromatogr. A*, 668:31 (1994).

14. C. A. Haynes, F. J. Benitez, H. W. Blanch, and J. M. Prausnitz, Application of integral-equation theory to aqueous two-phase partitioning systems, *AIChE J.*, 39:1539 (1993).

15. J. N. Baskir, T. A. Hatton, and U. W. Suter, Termodynamics of the partitioning of biomaterials in two-phase aqueous polymer systems: composition of lattice model to experimental data, *J. Phys. Chem.*, 93:2111 (1989).

16. D. R. Baughman, and Y. A. Liu, An expert network for extractive bioseparations in aqueous two-phase systems: an efficient strategy for experimental design and process development, American Chemical Society, San Diego, 1994.

17. V. P. Shanbhag, and C.-G. Axelsson, Hydrophobic interaction determined by partition in aqueous two-phase systems, *Eur. J. Biochem.*, 60:17 (1975).

18. M. A. Eiteman, and J. L. Gainer, Peptide hydrophobicity and partitioning in poly(ethylene glycol)/magnesium sulfate aqueous two-phase systems, *Biotechnol. Prog.* 6:479 (1990).

19. M. A. Eiteman, and J. L. Gainer, A model for the prediction of partition coefficients in aqueous two-phase systems, *Bioseparation*, 2:31 (1991).

20. P.-Å. Albertsson, A. Cajarville, D. E. Brooks, and F. Tjerneld, Partition of proteins in aqueous polymer two-phase systems and the effect of molecular weight of the polymer, *Biochim. Biophys. Acta* 926:87 (1987).

21. D. Forciniti, C. K. Hall, and M.-R. Kula, Protein partitioning at the isoelectric point: influence of polymer molecular weight and concentration and protein size, *Biotechnol. Bioeng.*, 38:986 (1991).

22. D. Forciniti, C. K. Hall, and M.-R. Kula, Influence of polymer molecular weight and temperature on phase composition in aqueous two-phase systems, *Fluid Phase Equil.*, 61:243 (1991).

23. P.-Å. Albertsson, "Partition of Cell Particles and Macromolecules," Wiley, New York (1986).

24. P.-Å. Albertsson, S. Sasakawa, and H. Walter, Cross partition and isoelectric points of proteins, *Nature*, 228:1329 (1970).

25. H. Walter, S. Sasakawa, and P.-Å. Albertsson, Cross-partition of proteins. Effect of ionic composition and concentration, *Biochemistry*, 11:3880 (1972).

26. C. L. DeLigny, and W. J. Gelsema, On the influence of pH and salt composition on the partition of polyelectrolytes in aqueous polymer two-phase systems, *Separ. Sci. Technol.*, 17:375 (1982).

27. M. A. Eiteman, and J. L. Gainer, Predicting partition coefficients in poly(ethylene glycol)/potassium phosphate aqueous two-phase systems, *J. Chromatog.*, 586:341 (1991).

28. M. A. Eiteman, and J. L. Gainer, The effect of the pH difference between phases on partitioning in poly(ethylene glycol)/phosphate aqueous two-phase systems, *Chem. Eng. Commun.*, 105:171 (1991).

29. M. A. Eiteman, Partitioning of charged solutes in poly(ethylene glycol)/potassium phosphate aqueous two-phase systems, *Sep. Sci. Technol.*, 29:685 (1994).

30. M. A. Eiteman, Predicting partition coefficients of multi-charged solutes in aqueous two-phase systems, *J. Chromatogr. A*, 668:21 (1994).

31. S. Bamberger, G. V. F. Seaman, J. A. Brown, and D. E. Brooks, The partition of sodium phosphate and sodium chloride in aqueous dextran poly(ethylene glycol) two-phase systems, *J. Colloid Inter. Sci.*, 99:187 (1984).

32. B. Yu. Zaslavsky, L. M. Miheeva, G. Z. Gasanova, and A. U. Mahmudov, Influence of inorganic electrolytes on partitioning of non-ionic solutes in an aqueous dextran-poly(ethylene glycol) biphasic system, *J. Chromatogr.*, 392:95 (1987).

33. O. Cascone, B. A. Andrews, and J. A. Asenjo, Partitioning and purification of thaumatin in aqueous two-phase systems, *Enzyme Microb. Technol.*, 13:629 (1991).

34. G. Johansson, Effects of different ions on the partitioning of proteins in an aqueous dextran-poly(ethylene glycol) two-phase system, *Proc. International Solvent Extraction Conf., Soc. of Chem. Ind., The Hague*, 2:928 (1971).

35. A. S. Schmidt, A. M. Ventom, and J. A. Asenjo, Partitioning and purification of α-amylase in aqueous two-phase systems, *Enzyme Microb. Technol.*, 16:131 (1994).

36. C.-K. Lee, and S. I. Sandler, Vancomycin partitioning in aqueous two-phase systems: effects of pH, salts, and an affinity ligand, *Biotechnol. Bioeng.*, 35:408 (1990).

37. A. D. Diamond, and J. T. Hsu, Fundamental Studies of Biomolecule Partitioning in Aqueous Two-Phase Systems, *Biotechnol. Bioeng.*, 34:1000 (1989).

38. R. F. Rekker, and H. M. de Kort, The hydrophobic fragmental constant; an extension to a 1000 data point set, *Eur. J. Med. Chem. Chim. Therapeut.*, 14:479 (1979).

39. M. A. Eiteman, C. Hassinen, and A. Veide, A mathematical model to predict the partition of peptides

and peptide-modified proteins in aqueous two-phase systems, *Biotechnol. Prog.*, in press (1994).

40. B. Skoog, Determination of polyethylene glycols 4000 and 6000 in plasma protein preparations. *Vox. Sang.*, 37:345 (1970).

41. Y. Nozaki, and C. Tanford, The solubility of amino acids and two glycine peptides in aqueous ethanol and dioxane solutions, *J. Biol. Chem.*, 246:2211 (1971).

TWO-PHASE AQUEOUS SURFACTANT SYSTEMS
FOR THE PURIFICATION OF BIOMATERIALS

C.-L. Liu, Y. J. Nikas, and D. Blankschtein

Department of Chemical Engineering and
Center for Materials Science and Engineering
Massachusetts Institute of Technology
Cambridge, MA 02139

We present a review of recent theoretical and experimental work aimed at investigating the potential use of two-phase aqueous surfactant systems for the purification of hydrophilic proteins by liquid-liquid extraction techniques. The systems studied include (1) a two-phase aqueous nonionic surfactant (n-decyl tetra(ethylene oxide), $C_{10}E_4$) system, and (2) a two-phase aqueous zwitterionic surfactant (dioctanoyl phosphatidylcholine, C_8-lecithin) system. The theoretical formulation assumes that excluded-volume interactions between the hydrophilic proteins and the surfactant micelles present in the solution play the dominant role in determining the experimentally observed partitioning behavior, and it incorporates (i) the self-assembling character of micelles, which allows them to grow into long, cylindrical microstructures with varying temperature and/or surfactant concentration, and (ii) a broad polydisperse distribution of micellar sizes. The theoretically predicted protein partitioning is compared with experimental measurements of the partitioning of several hydrophilic proteins, including cytochrome c, soybean trypsin inhibitor, ovalbumin, bovine serum albumin, and catalase, in two-phase aqueous $C_{10}E_4$ and C_8-lecithin systems, and is found to be in good agreement. The results of this investigation suggest that two-phase aqueous surfactant systems of the type considered in this paper are potentially useful as extractant phases for the liquid-liquid extraction of proteins and other biomaterials.

INTRODUCTION

In recent years, the utilization of two-phase aqueous surfactant systems to partition proteins and other biomaterials has received considerable attention.[1-9] Under the appropriate solution conditions, many surfactant systems can separate into two water-based, yet immiscible, liquid phases: one surfactant (micelle)-rich, and the other surfactant (micelle)-poor. Water-soluble proteins can partition unevenly between these two phases while maintaining

Aqueous Biphasic Separations: Biomolecules to Metal Ions
Edited by R.D. Rogers and M.A. Eiteman, Plenum Press, New York, 1995

49

their native conformations and biological activities. It is this feature that makes these systems particularly suitable for protein purification using liquid-liquid extraction techniques.

When compared to the more conventional two-phase aqueous *polymer* systems,[10-15] two-phase aqueous *surfactant* systems offer a number of unique, desirable features: (1) in the simplest realization, only a *binary* surfactant-water system is required instead of the more complex *ternary* polymer 1-polymer 2-water or polymer-salt-water systems, (2) the self-assembling, labile nature of the surfactant micelles present in the solution, as opposed to the unchanging identity of the polymers which is fixed upon synthesis, enables one to control and optimize the partitioning behavior by tuning micellar characteristics, including micelle shape and aggregation number, (3) the dual character of micelles, namely, that they can simultaneously offer hydrophilic and hydrophobic environments to solute species, gives rise to a partitioning selectivity based on the hydrophobicity of biomaterials, (4) the partitioning selectivity can be enhanced by utilizing mixed micelles containing surfactant-type ligands which can target a desired biomaterial, and (5) the separation of the desired biomaterial from the surfactant molecules after partitioning is completed can be facilitated by exploiting the self-assembling character of micelles, for example, micelles may be disassembled into their constituent surfactant monomers followed by filtration of the protein-surfactant monomer solution.

Considerable experimental work has been done[1-6] on the separation of hydrophilic (water-soluble) proteins from hydrophobic (water-insoluble) proteins using two-phase aqueous nonionic surfactant (Triton X-114) and zwitterionic surfactant systems. The separation in this case is based on the notion that hydrophobic proteins have a strong propensity to be incorporated (partially or totally) into micelles, and therefore should partition preferentially into the surfactant-rich phase in which there are many more micelles. On the other hand, hydrophilic proteins tend to remain in the aqueous environment, and therefore should partition preferentially into the surfactant-poor phase in which there is more available aqueous volume. In addition, the separation of hydrophilic proteins using ligand-containing mixed micelles has begun to be explored recently.[7]

In spite of these recent advances in the experimental arena, no theoretical work has been done to develop a fundamental understanding and quantitative description of protein partitioning in two-phase aqueous surfactant systems. As a first attempt to elucidate the mechanisms responsible for the observed partitioning behavior, we have recently developed[8] a theoretical description of the partitioning of hydrophilic proteins in two-phase aqueous surfactant systems containing non-charged micelles. The theory is based on the assumption that *excluded-volume interactions between the hydrophilic proteins and the non-charged micelles* play the dominant role in determining the experimentally observed partitioning behavior. The excluded-volume formulation incorporates (i) the self-assembling character of micelles, which allows them to change their shape and aggregation number with varying temperature and/or surfactant concentration, and (ii) a broad polydispersity in micellar sizes.

In this paper, we review our recent theoretical[8] and experimental work on protein partitioning in two-phase aqueous surfactant systems and present a comparison of the theoretically predicted and experimentally measured[9] partitioning behavior of several hydrophilic proteins of various sizes, including cytochrome *c*, soybean trypsin inhibitor, ovalbumin, bovine serum albumin, and catalase (in the order of increasing size), in the nonionic surfactant *n*-decyl tetra(ethylene oxide) $(C_{10}E_4)$-water and the zwitterionic surfactant dioctanoyl phosphatidylcholine $(C_8$-lecithin)-water two-phase systems. A detailed exposition of the theoretical and experimental work reviewed here, including a comparison of protein partitioning in two-phase aqueous surfactant and polymer systems, can be found in References 8 and 9.

MATERIALS AND EXPERIMENTAL METHODS

The nonionic surfactant $C_{10}E_4$ and the zwitterionic surfactant C_8-lecithin were selected for this study because (i) they are nondenaturing and gentle to proteins, and (ii) they form two-phase aqueous systems over a temperature range below typical protein denaturation temperatures.

Homogeneous $C_{10}E_4$ (lot no. 1006) was obtained from Nikko Chemicals (Tokyo). C_8-lecithin powder (lot no. 80PC-34) was obtained from Avanti Polar Lipids, Inc. (Alabaster, Alabama). Cytochrome c (from horse heart), soybean trypsin inhibitor (type I-S), ovalbumin, bovine serum albumin, and catalase (from bovine liver) were obtained from Sigma Chemicals (St. Louis, Missouri). All these materials were used as received. All other chemicals used were of reagent grade. All solutions were prepared using deionized water which had been fed through a Milli-Q ion-exchange system and were buffered at pH 7 by 10mM citric acid and 20mM disodium phosphate (McIlvaine buffer). Solutions also contained 0.02% sodium azide to prevent bacterial growth.

Coexistence curves for liquid-liquid phase separation of the buffered aqueous $C_{10}E_4$ and C_8-lecithin solutions without and with added proteins were determined by the cloud-point method.[8] This method consists of visually identifying the temperature, T_{cloud}, at which solutions of known surfactant concentrations become cloudy as the temperature is raised (for $C_{10}E_4$) or lowered (for C_8-lecithin). The measured cloud-point temperatures were reproducible to within 0.03°C.

The protein partition coefficient, K_p, provides a useful quantitative measure of the protein partitioning behavior, and is defined as the ratio of the protein concentration in the top phase, $C_{p,t}$, to that in the bottom phase, $C_{p,b}$, that is, $K_p = C_{p,t}/C_{p,b}$. For the measurement of K_p, buffered solutions containing known concentrations of either $C_{10}E_4$ or C_8-lecithin and protein were prepared and subsequently allowed to equilibrate at a constant temperature for at least 8 hours to form a two-phase system. The protein concentration in each coexisting phase was then determined by measuring the UV absorbance of that phase using a Shimadzu UV 160U spectrophotometer. Absorbance measurements were made at wavelengths of 545 nm for cytochrome c after it had been reduced by sodium ascorbate, 410 nm for catalase, and 280 nm for soybean trypsin inhibitor, ovalbumin, and bovine serum albumin, and were referenced to the absorbance of the identical $C_{10}E_4$ or C_8-lecithin solution phase (but without protein).

For more details on the experimental procedures for measuring coexistence curves and protein partition coefficients, see References 8 and 9.

EXPERIMENTAL RESULTS

Figure 1 shows the experimental coexistence curves of aqueous $C_{10}E_4$ surfactant solutions without protein ($\bigcirc$), and with 0.25 g/L cytochrome c ($\triangle$), 0.5 g/L ovalbumin ($*$), and 0.5 g/L catalase ($\square$) in McIlvaine buffer at pH 7. This figure indicates that over the range of surfactant concentrations examined, the added proteins have a negligible effect on the phase separation of the $C_{10}E_4$ solutions. Similar results were obtained for aqueous C_8-lecithin solutions containing proteins, but are not reported here due to space limitations. This important finding was utilized[8] in the theoretical formulation to decouple the description of the protein partitioning from that of the surfactant solution phase separation.

Figure 2 shows the experimentally determined partition coefficients, K_p, of cytochrome c ($\blacktriangle$), ovalbumin ($\bullet$), and catalase ($\blacksquare$) as a function of temperature over the range 18.8°C-21.2°C in two-phase aqueous $C_{10}E_4$ systems containing 0.25 g/L cytochrome *c*, 0.5 g/L ovalbumin, and 0.5 g/L catalase, respectively, in McIlvaine buffer at pH 7. The fact that $K_p < 1$ indicates that these three hydrophilic proteins partition preferentially into the *bottom micelle-poor phase*. It is also clear that, as the temperature increases away from the critical temperature, $T_c \approx 18.8°C$ (corresponding to the minimum of the coexistence curve in Figure 1), K_p decreases and deviates further from unity for all the three proteins. These observations suggest that (i) proteins are pushed into that phase which has a larger available free volume (which, in this case, is the bottom micelle-poor phase) due to

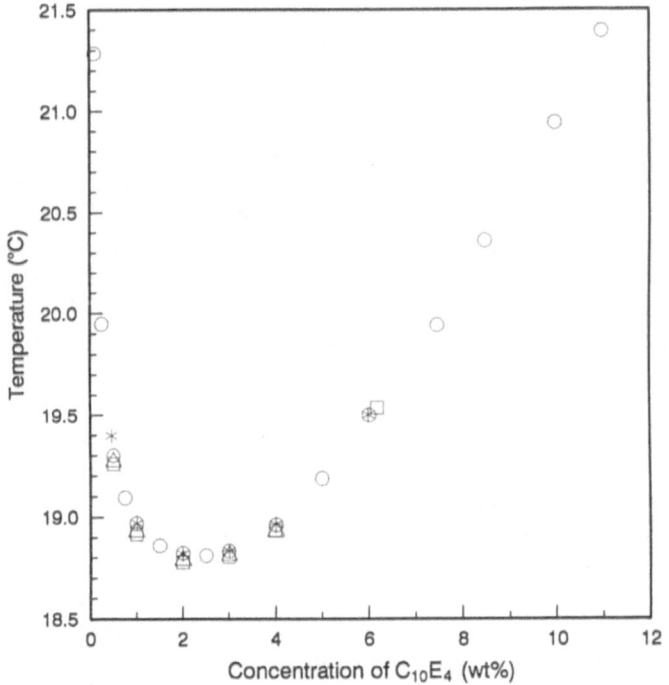

Figure 1. Experimentally measured coexistence curves of aqueous solutions of $C_{10}E_4$ without protein (○) and with 0.25 g/L cytochrome *c* (△), 0.5 g/L ovalbumin (∗), and 0.5 g/L catalase (□). The area above the curve is the two-phase region, in which the partitioning experiments were conducted.

excluded-volume interactions between $C_{10}E_4$ micelles and protein molecules, and (ii) this tendency becomes stronger as $T - T_c$ increases, that is, with increasing difference in the surfactant concentrations (or the volume fractions occupied by micelles) of the two coexisting phases (see Figure 1). One can also observe from Figure 2 that, at a fixed temperature, the extent of the protein partitioning into the bottom micelle-poor phase increases in the order cytochrome *c* < ovalbumin < catalase. This observed trend is consistent with the notion that excluded-volume interactions between proteins and $C_{10}E_4$ micelles play the dominant role in determining the observed partitioning behavior, since catalase has the largest size (M.W. 232 000 Da), followed by ovalbumin (44 000 Da), and cytochrome *c* (12 400 Da).

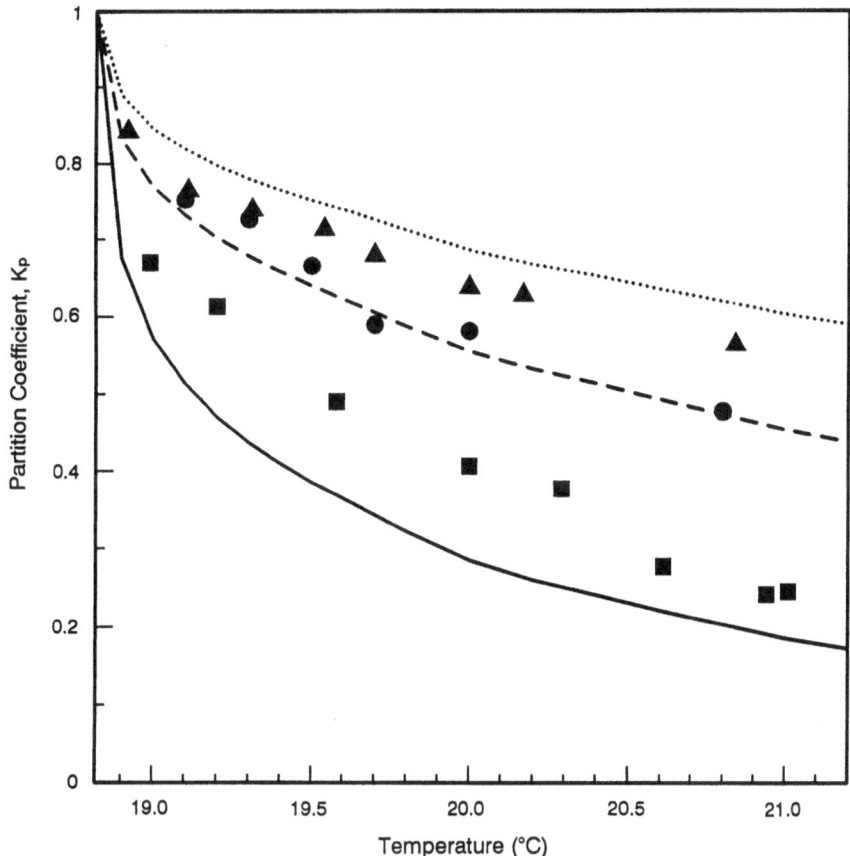

Figure 2. Experimentally measured partition coefficients, K_p, of cytochrome c (▲), ovalbumin (●), and catalase (■) at different temperatures in the two-phase aqueous $C_{10}E_4$ surfactant system. Also shown are the predicted partition coefficients K_p of cytochrome c ($\cdots$), ovalbumin ($--$), and catalase (—) as a function of temperature.

THEORETICAL FORMULATION

Results of various experimental and theoretical studies have shown[16-18] that, under appropriate solution conditions, $C_{10}E_4$ and C_8-lecithin can form long, flexible, and polydisperse cylindrical micelles. The average length of a $C_{10}E_4$ or C_8-lecithin micelle greatly exceeds the hydrodynamic radius of a typical hydrophilic protein (20Å-50Å), and the micelle is rigid on the scale of a typical protein molecule.[8] Experimental results also indicate[19,20] that hydrophilic proteins do not bind nonionic and zwitterionic surfactants of the type used in this study to any significant extent. In addition, our coexistence curve measurements indicate that the effect of these proteins on micellar characteristics is negligible. In view of this, it is reasonable to assume that, to a first approximation, hydrophilic proteins and $C_{10}E_4$ or C_8-lecithin micelles behave as mutually non-associating entities interacting primarily through short-ranged, repulsive, excluded-volume interactions.

Specifically, we have assumed[8] that an aqueous surfactant-protein solution containing non-charged cylindrical micelles and globular hydrophilic proteins can be modeled as a mixture of mutually penetrable, polydisperse, hard spherocylinders (micelles) and hard spheres (proteins).

As stated earlier, the protein partition coefficient is defined as

$$K_p = C_{p,t}/C_{p,b} \tag{1}$$

where $C_{p,t}$ and $C_{p,b}$ are the protein concentrations in the top (t) and bottom (b) phases respectively. Under conditions of low protein concentration, uncharged surfactants, and low salt concentration, K_p is given by[10]

$$K_p = \exp\left[-(\mu_{p,t}^0 - \mu_{p,b}^0)/k_B T\right] \tag{2}$$

where $\mu_{p,t}^0$ and $\mu_{p,b}^0$ are the standard-state chemical potentials of a protein molecule in the top and bottom phases respectively, k_B is the Boltzmann constant, and T is the absolute temperature.

In a system comprising hard particles, the chemical potentials are determined solely by entropic factors, and the difference between the standard-state chemical potentials of a protein molecule in the top and bottom phases is given by[21]

$$\mu_{p,t}^0 - \mu_{p,b}^0 = -k_B T \ln(\Omega_t/\Omega_b) \tag{3}$$

where Ω_t and Ω_b denote the numbers of ways of placing a protein molecule in the top and bottom phases respectively. Combining Eqs. (2) and (3), we obtain the following simple expression for the protein partition coefficient

$$K_p = \Omega_t/\Omega_b \tag{4}$$

Under conditions that (i) the number densities of micelles in the top and bottom phases are very low, and (ii) there is no spatial or orientational order in the distribution of the micelles, the number of ways, Ω, of placing a protein molecule in an isotropic and homogeneous solution of volume V containing a distribution $\{N_n\}$ of micelles having aggregation number n ($n = 1, ...\infty$) is given by[8]

$$\Omega = A \exp\left(-\sum_n N_n U_{n,p}/V\right) \tag{5}$$

where N_n is the number of micelles of aggregation number n, $U_{n,p}$ is the excluded volume between a micelle of aggregation number n and a protein molecule, and A is a proportionality constant.

The excluded volume, $U_{n,p}$, between a protein molecule (sphere) and a micelle (spherocylinder with hemispherical ends) is given by

$$U_{n,p} = 4\pi(R_n + R_p)^3/3 + \pi(R_n + R_p)^2 L_n \tag{6}$$

where R_n and L_n are the cross-sectional radius and length of the cylindrical part of a spherocylindrical micelle of aggregation number n, and R_p is the radius of the protein molecule. The value of R_n is determined by the "length" of the constituent surfactant monomers, and is therefore independent of the micellar aggregation number, n. We can therefore write $R_n = R_0$. On the other hand, the value of L_n increases linearly with n and is related to the total volume of the micelle, nv_s, (where v_s is the volume that a surfactant monomer occupies in the micelle, which *may not* be the same as the actual physical volume of the surfactant molecule, see the next section) by

$$\pi R_0^2 L_n = nv_s - 4\pi R_0^3/3 \tag{7}$$

where $4\pi R_0^3/3$ is the volume of the two hemispherical caps of a spherocylindrical micelle. Using Eq. (7) in Eq. (6), we obtain the following expression for $U_{n,p}$

$$U_{n,p} = nv_s(1 + R_p/R_0)^2 + V_p(1 + R_0/R_p)^2 \tag{8}$$

where $V_p = 4\pi R_p^3/3$ is the volume of the protein molecule. Substituting the expression for $U_{n,p}$ given in Eq. (8) in Eq. (5) and carrying out the summation yields

$$\Omega = A \exp\left\{-\left[(1 + R_0/R_p)^2 \rho V_p + (1 + R_p/R_0)^2 \phi\right]\right\} \tag{9}$$

where $\rho = \sum_n N_n/V$ is the number density of micelles, and $\phi = N_s v_s/V$ is the total volume fraction occupied by micelles, with $N_s = \sum_n n N_n$ being the total number of surfactant molecules. (Note that the surfactant monomer concentration is ignored here because it is very close to the critical micelle concentration and hence nearly equal in the top and bottom micellar solution phases. Therefore, the presence of the monomers has little effect on the protein partitioning behavior.) Note that ϕ represents the fraction of the solution volume which is actually occupied by micelles, and it depends on the volume occupied by a surfactant molecule in a micelle, v_s. Consequently, ϕ *may not* correspond to the actual total volume fraction of surfactant (see the next section for details).

The number density of micelles, ρ, in Eq. (9) is determined by the size distribution, $\{N_n\}$, of the micelles. When the micelles grow into long cylinders (which is the case of interest here), ρV_p can be shown[8] to be much smaller than ϕ. In that case, Ω in Eq. (9) is given approximately by

$$\Omega = A \exp\left[-\phi(1 + R_p/R_0)^2\right] \tag{10}$$

Using Eq. (10) for the top (t) and bottom (b) phases respectively in Eq. (4), one obtains the following remarkably simple expression for the protein partition coefficient in a two-phase aqueous surfactant system containing long cylindrical micelles

$$K_p = \exp\left[-(\phi_t - \phi_b)(1 + R_p/R_0)^2\right] \tag{11}$$

where ϕ_t and ϕ_b denote the *volume fractions occupied by micelles* in the top and bottom phases respectively.

Equation (11) indicates that the uneven partitioning of a hydrophilic protein in the two-phase aqueous non-charged surfactant systems considered here is a direct consequence of the difference in the volume fractions that micelles occupy in the two coexisting micellar solution phases, $(\phi_t - \phi_b)$. In addition, the value of the partition coefficient depends on the relative sizes of micelles and proteins, as reflected in the values of R_0 and R_p. Specifically, as the protein size increases, the protein will partition more unevenly into the micelle-poor phase of the two-phase aqueous surfactant system.

THEORETICAL PREDICTIONS AND COMPARISON WITH EXPERIMENTS

In order to predict the variation of K_p with temperature, values of R_0 and R_p and of $(\phi_t - \phi_b)$ as a function of temperature are needed. In general, R_0 is approximately equal to the length of the surfactant molecule in a micelle, and can be written as the sum of the cross-sectional radius of the hydrocarbon core, l_c, and the length of the surfactant hydrophilic moiety (referred to as "head"), l_h, that is, $R_0 = l_c + l_h$.

Calculations based on a recently developed molecular model of micellization[17,22] yield $l_c \approx 12$Å and 10Å for $C_{10}E_4$ and C_8-lecithin micelles respectively. The value of l_h depends on the average conformation adopted by the surfactant head, which is a tetra(ethylene oxide) chain in the case of $C_{10}E_4$ and a phosphatidylcholine group in the case of C_8-lecithin. As a first approximation, we assume that the unperturbed tetra(ethylene oxide) chains of $C_{10}E_4$ micelles behave as Gaussian chains with one end attached to a wall (to mimic the micellar surface). This results in a value of $l_h \approx 9$Å.[23] The phosphatidylcholine group in a C_8-lecithin micelle is assumed to be fully extended and oriented perpendicular to the micelle surface, in which case its length is estimated to be 11Å. Therefore, the cross-sectional radii of $C_{10}E_4$ and C_8-lecithin micelles are both approximately given by $R_0 = 21$Å. The hydrodynamic radii of cytochrome c, ovalbumin, and catalase are $R_p = 19$Å, 29Å, and 52Å, respectively.[14]

As stated earlier, the volume fractions of micelles in the top and bottom phases, ϕ_t and ϕ_b, in the case of $C_{10}E_4$ micelles, can be different from the total volume fractions of surfactants, ϕ_t' and ϕ_b', as determined from Figure 1. This is due to the substantial water penetration into the region containing surfactant heads when the tetra(ethylene oxide) chains adopt a Gaussian conformation. In this case, the "wet" volume occupied by a surfactant molecule in a micelle, v_s, can be larger than the actual "dry" volume of a surfactant molecule, v_s'. When there is no water penetration into the region containing the heads, the length of the "dry" surfactant head in a cylindrical micelle, l_h', is given by $l_h' = l_c[(v_s'/v_c)^{1/2} - 1]$, where v_s' and v_c are the actual "dry" volumes of the surfactant molecule and its hydrocarbon tail respectively. For $C_{10}E_4$ micelles, $v_s' = 580$Å^3, and $v_c = 269$Å^3 (from Reference 8), which yields $l_h' = 5.8$Å.

When the head region is highly hydrated, such that $l_h > l_h'$, the volume of a cylindrical micelle is greater than the total physical "dry" volume occupied by the constituent surfactant molecules, and should be therefore scaled by a correction factor $[(l_c + l_h)/(l_c + l_h')]^2$. Accordingly, ϕ_t (ϕ_b) is equal to ϕ_t' (ϕ_b') multiplied by the correction factor given above.

In the case of C_8-lecithin micelles, the heads are more compact than in the $C_{10}E_4$ case, and, therefore, the extent of water penetration is expected to be less pronounced. Accordingly, as an approximation, we assume that the volume of a C_8-lecithin micelle is equal to the total physical "dry" volume occupied by the surfactant molecules constituting the micelle. Consequently, the correction factor is taken as 1 for C_8-lecithin micelles.

The values of ϕ_t' and ϕ_b' at various temperatures can be obtained from the experimentally measured coexistence curves of aqueous solutions of $C_{10}E_4$ or C_8-lecithin. As an illustration, we consider the coexistence curve corresponding to $C_{10}E_4$, which is shown in Figure 1. At a given temperature, ϕ_t' and ϕ_b' are given by the intersections of the horizontal tie line corresponding to that temperature with the surfactant-rich and surfactant-poor branches of the coexistence curve respectively. Using the known values of l_c, l_h, and l_h' given earlier, the correction factor for $C_{10}E_4$ can be calculated (recall that it is unity for C_8-lecithin), which when multiplied by $(\phi_t' - \phi_b')$ yields the values of $(\phi_t - \phi_b)$ to be used in Eq. (11).

Figure 2 shows the predicted variation of K_p temperature (in the two-phase aqueous $C_{10}E_4$ system) for cytochrome c ($\cdots$), ovalbumin ($--$), and catalase ($—$) corresponding to the $(\phi'_t - \phi'_b)$ values determined from Figure 1, $R_0 \approx 21\text{Å}$, and the R_p values listed above. As can be seen, there is good agreement with the experimentally measured K_p values.

The dependence of the partition coefficient, K_p, on protein size, R_p, can be seen clearly by plotting K_p as a function of the ratio R_p/R_0, at a fixed temperature, or equivalently, at a fixed value of $(\phi'_t - \phi'_b)$. Specifically, for $C_{10}E_4$ at 21°C, $\phi'_t - \phi'_b \approx 10\%$ (see

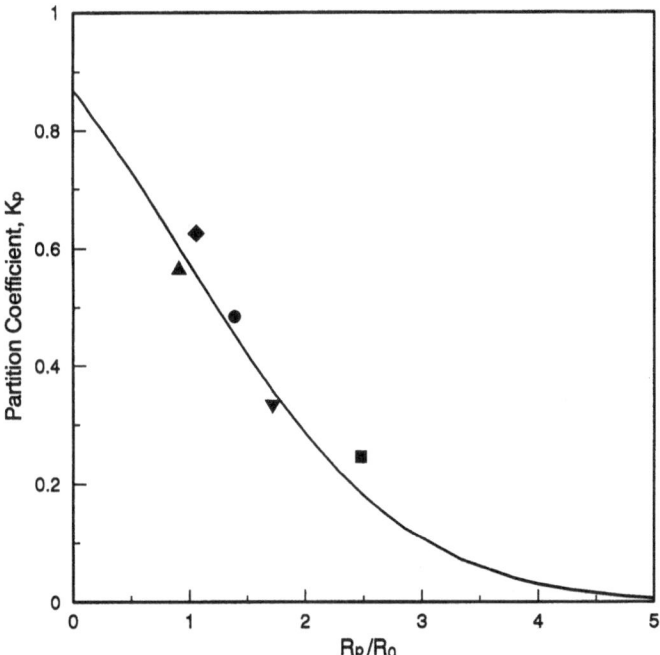

Figure 3. Predicted protein partition coefficient, K_p, as a function of the ratio R_p/R_0 in aqueous $C_{10}E_4$ surfactant solutions. R_p is the protein hydrodynamic radius, $R_0 = 21\text{Å}$ is the cross-sectional radius of a $C_{10}E_4$ cylindrical micelle, and $\phi'_t - \phi'_b = 10\%$. The various symbols correspond to the experimentally measured K_p values of the following proteins: ($\blacktriangle$, $R_p=19\text{Å}$) cytochrome c, ($\blacklozenge$, $R_p=22\text{Å}$) soybean trypsin inhibitor, ($\bullet$, $R_p=29\text{Å}$) ovalbumin, ($\blacktriangledown$, $R_p=36\text{Å}$) bovine serum albumin, and ($\blacksquare$, $R_p=52\text{Å}$) catalase.

Figure 1). Figure 3 shows the predicted variation of K_p as a function of R_p/R_0 (full line), together with the experimental K_p values corresponding to cytochrome c ($\blacktriangle$), soybean trypsin inhibitor ($\blacklozenge$), ovalbumin ($\bullet$), bovine serum albumin ($\blacktriangledown$), and catalase ($\blacksquare$). The hydrodynamic radii of soybean trypsin inhibitor and bovine serum albumin are $R_p=22\text{Å}$ and 36Å respectively.[24] This figure indicates that as R_p increases relative to R_0, the value of K_p decreases and can become vanishingly small for $R_p/R_0 > 5$.

In the case of C_8-lecithin, for illustrative purposes, we selected a temperature of 10°C, at which $(\phi_b' - \phi_t') \approx 10\%$. Note that in the C_8-lecithin case, *the bottom phase is micelle-rich while the top phase is micelle-poor*. Accordingly, due to excluded-volume interactions, hydrophilic proteins should partition preferentially into the top micelle-poor phase, namely, the values of K_p should be greater than 1. Figure 4 shows the predicted

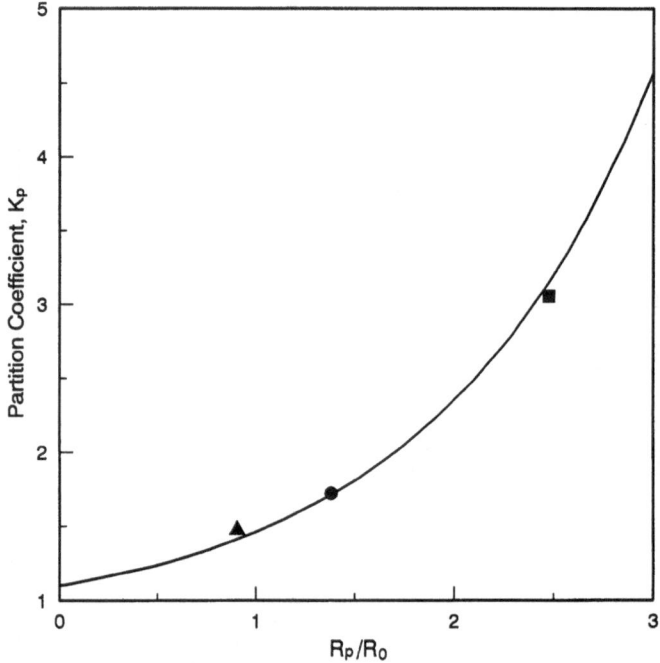

Figure 4. Predicted protein partition coefficient, K_p, as a function of the ratio R_p/R_0 in aqueous C_8-lecithin surfactant solutions. R_p is the protein hydrodynamic radius, R_0=21Å is the cross-sectional radius of a C_8-lecithin cylindrical micelle, and $\phi_b' - \phi_t' = 10\%$. The various symbols correspond to the experimentally measured K_p values of the following proteins: (▲, R_p=19Å) cytochrome c, (●, R_p=29Å) ovalbumin, and (■, R_p=52Å) catalase.

variation of K_p as a function of R_p/R_0 using Eq. (11), together with the experimental K_p values corresponding to cytochrome c (▲), ovalbumin (●), and catalase (■). This figure shows that, as expected, $K_p > 1$ and increases as R_p/R_0 increases. One can see from Figures 3 and 4 that the agreement between theory and experiment is good for both surfactant systems.

SUMMARY

We have reviewed some recent theoretical and experimental work on protein partitioning in two-phase aqueous $C_{10}E_4$ and C_8-lecithin surfactant systems. The good agreement between theory and experiment indicates that excluded-volume interactions between non-charged micelles and proteins are the dominant factor in determining the observed protein partitioning behavior, as captured in our theoretical formulation. Our results suggest that this type of two-phase aqueous surfactant systems should be particularly useful to partition large hydrophilic particles, such as water-soluble colloids and cells.

ACKNOWLEDGMENTS

This research was supported in part by the National Science Foundation (NSF) Presidential Young Investigator (PYI) Award to Daniel Blankschtein, and an NSF Grant No. DMR-84-18778 administered by the Center for Materials Science and Engineering at M.I.T. Daniel Blankschtein is also grateful to BASF, Kodak, Intevep, S.A., and Unilever for partial support of this project.

REFERENCES

1. C. Bordier, Phase separation of integral membrane proteins in Triton X-114 solution, *J. Biol. Chem.*, 256:1604 (1981).

2. J.G. Pryde, Triton X-114: a detergent that has come in from the cold, *TIBS*, 5:160 (1986).

3. J.G. Pryde and J.H. Phillips, Fractionation of membrane proteins by temperature-induced phase separation in Triton X-114, *Biochem. J.*, 233:525 (1986).

4. C. Holm, G. Fredrikson, and P. Belfrage, Demonstration of the amphiphilic character of hormone-sensitive lipase by temperature-induced phase separation in Triton X-114 and charge-shift electrophoresis, *J. Biol. Chem.*, 261:15659 (1986).

5. R.A. Ramelmeier, G.C. Terstappen, and M.-R. Kula, The partitioning of cholesterol oxidase in Triton X-114-based aqueous two-phase systems, *Bioseparation*, 2:315 (1991).

6. T. Saitoh and W.L. Hinze, Concentration of hydrophobic organic compounds and extraction of protein using alkylammoniosulfate zwitterionic surfactant mediated phase separations (cloud point extractions), *Anal. Chem.*, 63:2520 (1991).

7. T. Saitoh and W.L. Hinze, Use of surfactant-mediated phase separation (cloud point extraction) with affinity ligands for the extraction of hydrophilic proteins, Preprint (1992).

8. Y.J. Nikas, C.-L. Liu, T. Srivastava, N.L. Abbott, and D. Blankschtein, Protein partitioning in two-phase aqueous nonionic micellar solutions, *Macromolecules*, 25:4794 (1992).

9. C.-L. Liu, Y.J. Nikas, and D. Blankschtein, Partitioning of proteins using two-phase aqueous surfactant systems, *AIChE J.*, in press (1994).

10. P.-Å. Albertsson, *Partition of Cell Particles and Macromolecules*, Wiley-Interscience, New York (1986).

11. H. Walter, D.E. Brooks, and D. Fisher, ed., *Partitioning in Aqueous Two-Phase Systems: Theory, Methods, Uses, and Applications to Biotechnology*, Academic Press, Orlando (1985).

12. D. Fisher and I.A. Sutherland, ed., *Separations Using Aqueous Phase Systems: Applications in Cell Biology and Biotechnology*, Plenum Press, New York (1989).

13. H. Walter, G. Johansson, and D.E. Brooks, Partitioning in aqueous two-phase systems: recent results, *Anal. Biochem.*, 197:1 (1991).

14. N.L. Abbott, D. Blankschtein, and T.A. Hatton, On protein partitioning in two-phase aqueous polymer systems, *Bioseparation*, 1:191 (1990).

15. N.L. Abbott, D. Blankschtein, and T.A. Hatton, Protein partitioning in two-phase aqueous polymer systems. 1. Novel physical pictures and a scaling-thermodynamic formulation, *Macromolecules*, 24:4334 (1991); 2. On the free energy of mixing globular colloids and flexible polymers, *Macromolecules*, 25:3917 (1992); 3. A neutron scattering investigation of the polymer solution structure and protein-polymer interactions, *Macromolecules*, 25:3932 (1992); 4. Proteins in solutions of entangled polymers, *Macromolecules*, 25:5192 (1992); 5. Decoupling of the effects of protein concentration, salt type, and polymer molecular weight, *Macromolecules*, 26:825 (1993).

16. D. Blankschtein, G.M. Thurston, and G.B. Benedek, Phenomenological theory of equilibrium thermodynamic properties and phase separation of micellar solutions, *J. Chem. Phys.*, 85:7268 (1986).

17. S. Puvvada and D. Blankschtein, Molecular-thermodynamic approach to predict micellization, phase behavior and phase separation of micellar solutions. I. Application to nonionic surfactants, *J. Chem. Phys.*, 92:3710 (1990).

18. B. Lindman and H. Wennerström, Nonionic micelles grow with increasing temperature, *J. Phys. Chem.*, 95:6053 (1991).

19. A. Helenius and K. Simons, The binding of detergents to lipophilic and hydrophilic proteins, *J. Biol. Chem.*, 247:3656 (1972).

20. S. Makino, J.A. Reynolds, and C. Tanford, The binding of deoxycholate and Triton X-100 to proteins, *J. Biol. Chem.*, 248:4926 (1973).

21. L.D. Landau and E.M. Lifshitz, *Statistical Physics*, 2nd ed., Vol. 1, Pergamon Press, Oxford (1980).

22. A. Naor, S. Puvvada, and D. Blankschtein, An analytical expression for the free energy of micellization, *J. Phys. Chem.*, 96:7830 (1992).

23. C. Sarmoria and D. Blankschtein, Conformational characteristics of short poly(ethylene oxide) chains terminally attached to a wall and free in aqueous solution, *J. Phys. Chem.*, 96:1978 (1992).

24. P.L. Dubin and J.M. Principi, Optimization of size-exclusion separation of proteins on a Superose column, *J. Chromatography*, 479:159 (1989).

INTEGRATION OF AQUEOUS TWO-PHASE EXTRACTION WITH OTHER SEPARATION TECHNIQUES

Rajni Kaul and Bo Mattiasson

Department of Biotechnology
Chemical Center, Lund University
Box 124, S-221 00 Lund, Sweden

INTRODUCTION

Downstream processing constitutes a major part of the costs in production processes of proteins. In most cases a whole sequence of separation steps are applied before a pure product is obtained. Each step involves losses, and therefore it is of interest to reduce the number of steps without losing the quality of the product. There has been a trend in fermentation technology to integrate the conversion step with one or two product separation steps with an aim to achieve `in situ´ product recovery and also to improve the productivity of the processes.[1,2] Integration of different separation techniques is a viable concept even with respect to protein purification procedures. The ultimate goal of this integration is to achieve a combination of efficiency, selectivity and favorable scale up characteristics.[3]

Extraction in aqueous two-phase systems is a separation technique that is amenable to easy scale up. It has proven to be a very useful tool for rapid isolation of proteins from cells and cell debris, and even soluble material without the use of high speed centrifugation, etc. The technique is, therefore, regarded as suitable for application in the early stages of a separation train. In a typical extraction, the conditions are such that the

Aqueous Biphasic Separations: Biomolecules to Metal Ions
Edited by R.D. Rogers and M.A. Eiteman, Plenum Press, New York, 1995

61

particulate matter and some other contaminants partition to the bottom phase while the protein of interest goes to the top phase. PEG/salt systems have been commonly used. The extraction may be continued further to remove the protein from the top phase into a new salt phase. This concept has been successfully used to isolate a number of proteins from crude homogenates; in several cases leading to substantial purification of the protein product.[4]

Despite the simplicity of the technique, use of aqueous two-phase systems has not become as common in industry as one should imagine after having evaluated the literature from academic laboratories. There are still a few drawbacks that have hampered the development; e.g. 1) the difficulties to predict the partition behaviour, 2) the cost of the phase forming components and waste water treatment, and 3) presence of traces of polymers in the final product.

As the spontaneous partitioning of biomolecules in two-phase systems is a combination of different factors, one has to invest in laborious and time consuming work to find the optimal phase system for each separation need. In many cases, desired partitioning may be difficult to achieve. Extraction of proteins has been made more selective and predictable by affinity mediated partitioning.[5] For this, affinity ligands specific for the desired target protein are introduced into the two-phase system by way of coupling to one of the phase forming polymers, usually PEG, thus ensuring its residence in that particular phase. But this approach has also not been feasible on large scale mainly because of the other limitations of the aqueous two-phase extraction mentioned above. The costs of the polymers has resulted in research towards alternative, less expensive polymers.[6] A parallel line of research has been to develop efficient procedures for recovery and recycling of the phase forming chemicals, which would, besides lowering the process costs, reduce the load for waste water treatment.[7] Traces of polymers in the final product may be removed through subsequent steps e.g. by specific adsorption of the protein on a chromatography matrix.

INTEGRATING EXTRACTION WITH OTHER SEPARATIONS

A few strategies for facilitating the efficiency of aqueous two-phase separation have been attempted by means of incorporating other separation materials. In Table 1 are listed some approaches that have been tried, including the advantages gained by these means. Besides reducing the number of steps, one can also recognize that in some cases a more efficient/predictable separation takes place. This is valid both for the phase separation as such, and in some cases it promotes the partitioning of the target molecule in the form of an affinity complex. Still another advantage may be that the further downstream processing is facilitated by appropriate choice of reagents in the primary extraction step.

Table 1. Examples of process integration in downstream processing involving an extraction step in aqueous two-phase systems.

Separation method	System studied	Comment	Reference
Magnetic separation	Different two-phase systems	improves efficiency of separation	8, 9
Chromatography	Protein A-human gamma globulin BSA, ADH, pyruvate kinase-Cibacron blue	good prediction of partition behaviour of ligand, better control of amount of active ligand; good elution conditions, product free of traces of phase polymer	10-12
Precipitation	Protein A-IgG LDH-Cibacron blue	good elution conditions, easy removal of traces of phase polymer from product	13 14

The way the different methods could be combined in order to obtain an optimal process configuration is not yet obvious and would depend on the system under consideration.

Magnetic field facilitated phase separation

It is known that aqueous two-phase systems are convenient to work with, since they are easy to mix and that they separate spontaneously into two phases within reasonable time periods, often 10-20 minutes. Addition of large amounts of proteins or complex material such as a cell homogenate to a phase system may substantially increase the separation time due to surface active properties of the biomaterial. Faster separation is often achieved by centrifugation. An alternative that seems very convenient is the addition of magnetically susceptible material e.g. iron oxide particles to the phase system so that the systems separate rapidly, in a matter of seconds, when an electromagnetic field is applied.[8,9] By such an approach it may also be possible to utilize phase systems that otherwise are rejected because of their relative sluggishness with regard to phase separation, e.g., high viscosity systems or phases with similar densities.

Affinity extraction x chromatography

Solid chromatographic particles are used as affinity adsorbents in the aqueous two-phase system for extraction of proteins. Partition behaviour of the particles is easier to be predicted and their separation from the phase system for further processing is also simpler implying a straightforward ligand recovery and reuse.[10-12] Various forms of Sepharose particles have been found to distribute quantitatively to the PEG phase and then settle on the interface after centrifugation, whereas the other commercially available chromatographic resins partition to the bottom phase.[10,12] In the studies reported, the system could be further strengthened by modifying the Sepharose particles with a phase polymer (PEG) so as to obtain exclusive partitioning to the desired phase. The ligand bearing particles may be mixed with the crude homogenate prior to partitioning in the two-phase system. After phase separation has taken place, the cell debris accumulates in the bottom phase; the particles are harvested, from the interface, placed in a column, washed and the bound protein eluted according to traditional chromatographic methods.[10,11] The high salt concentration normally used in conventional two-phase extraction is thus avoided. On the other hand, it is common knowledge that conventional affinity chromatography, though highly resolving, requires the sample to be clean and free of any particulate matter, and is normally performed towards the end of a purification scheme. Hence, the integration of chromatography and aqueous two-phase systems combines the advantages of extraction with the simple washing and elution conditions provided by chromatography.

Affinity extraction x precipitation

Precipitation is a separation technique that is routinely utilized in biotechnology industry on large scale. Integration of precipitation with aqueous two-phase extraction makes use of the precipitability of the polymers. This provides a means to effect their removal from the final product and makes their recycling easy. Recently, polymers having low cloud point e.g. UCON, have been used as one of the phase forming components in two-phase systems.[15] The polymer is simply heated up to a temperature above its cloud point for it to separate into a polymer phase and the aqueous phase containing the biomolecules, etc. Employing such a polymer even as a ligand carrier has been shown to facilitate the final recovery of the protein from the affinity complex.[16]

A variety of polymers having a property of reversible solubility-insolubility with moderate change in an environmental parameter such as pH, temperature, ionic strength etc. are known.[3] Such polymers have been used as carriers of ligand molecules in a technique called affinity precipitation, where the soluble form of the modified ligand is used during affinity binding, and precipitation is induced for achieving separation. Introduction of such polymers as ligand carriers in aqueous two-phase systems provides an interesting system with features of quick removal of cell debris, separate handling of the phase polymer and the modified ligand, etc. Some of the studies carried out in our laboratory using this approach are briefly described here.

Eudragit as a ligand carrier in aqueous two-phase systems. Eudragit S 100, a copolymer of methacrylate and methyl methacrylate, is an example of a polymer which is made reversibly soluble and insoluble with changes in pH. The polymer has a molecular weight of 135 000, and possesses an ester to carboxyl group ratio of 2:1. A number of affinity ligands have been bound to Eudragit using carbodiimide coupling procedure. Figure 1 shows the precipitation profile of Eudragit and the polymer with the bound ligand with respect to change in pH. The polymer has a sharp soluble-insoluble transition profile between pH 4.5 and 5.5. It has been observed that binding of any ligand to Eudragit results in the shifting of the transition curve slightly upward on the pH scale, making it possible to precipitate the polymer-ligand complex at relatively high pH value.

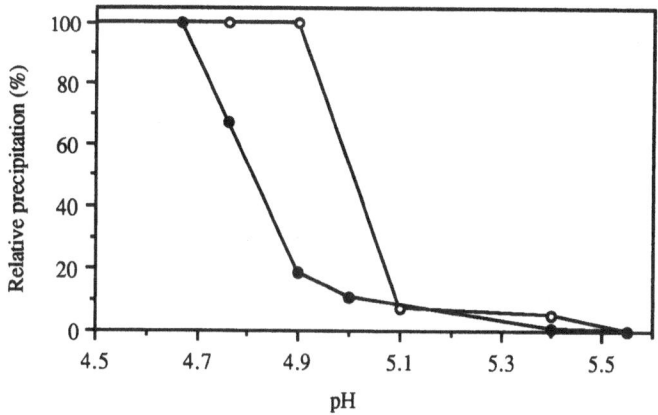

Figure 1. Precipitation profile of Eudragit S 100 (●) and Eudragit with bound ligand (○). The precipitation is followed by measuring the turbidity at 470 nm.

Figure 2 shows the partitioning of the polymer in a PEG 8000-Dextran 250 system. At neutral pH, the polymer is equally distributed in the two phases, but the partitioning is changed depending on the kind of salt anions present. Low concentration of phosphate make Eudragit partition preferentially to the top phase. On the other hand, NaCl has an opposite effect on the partitioning. It is perhaps the electrostatic repulsion between Eudragit, a polyanion and the salt anions that generates the partitioning. In the PEG/Reppal system, the polymer favors the PEG phase, and addition of low concentration of phosphate makes the partitioning more complete in top phase.[13] In the absence of the salt ions, the partitioning of Eudragit could be influenced by the pH of the system, the polymer moving up to the PEG phase at lower pH values in a PEG/Dextran system. This could be attributed to the predominantly hydrophobic character of Eudragit as the pH is reduced.

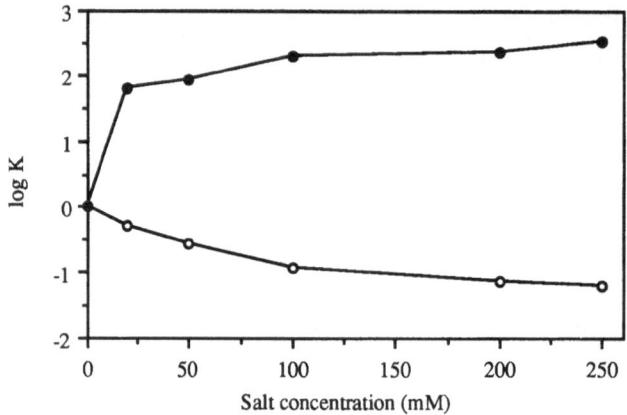

Figure 2. Partitioning of Eudragit S 100 in 6% (w/w) PEG 8000 - 8% (w/w) Dextran 250 system in the presence of potassium phosphate (●) and sodium chloride (○).

The partitioning of Eudragit bound ligands to the upper phase sets a stage for the extraction of the target protein as in conventional affinity extraction. The strategy is then to precipitate the polymer bound affinity complex from the upper phase by lowering the pH, and then treat the precipitate for desorption of the bound protein. Thereafter, the Eudragit-ligand is recycled.

Using the above approach, the first study reported was that of Eudragit bound immunoglobulin G (IgG) for the purification of Protein A from recombinant *Escherichia coli.*. The crude cell homogenate was mixed directly with the phase components, giving a final concentration of 5% (w/w) PEG 8000, 14% (w/w) Reppal PES 200, 0.5% (w/w) Eudragit-IgG and 0.2 M potassium phosphate. A preparation having 60 mg IgG immobilized per gm Eudragit was used having an adsorption capacity of about 4 mg of Protein A. After phase equilibration and separation, the bottom phase containing the cell debris was separated. From the top phase the Eudragit-IgG-Protein A complex was precipitated by reducing the pH to 4.5, and separated by centrifugation at 4500 rpm for 10 min. After washing with 0.01 M acetate buffer, pH 4.5, the Protein A was desorbed by treatment twice with 0.1 M glycine-HCl buffer, pH 2.5. The results of the purification of Protein A using this procedure are presented in Table 2. Best product recovery (85%) and purification (26 fold) were obtained when the added sample contained Protein A amount equivalent to the binding capacity of Eudragit-IgG. Overloading the homogenate resulted in excess Protein A remaining in the bottom phase.

The relative adsorption capacity of Eudragit-IgG was slightly reduced after the first use and thereafter remained constant. However, the amount of protein accumulated on Eudargit-IgG increased with time. This is to be expected because of the nature of the polymer backbone which allows non-specific adsorption of biomolecules via charge-charge and/or hydrophobic interactions.

Another system studied was the affinity extraction of lactate dehydrogenase from porcine muscle extract using Eudragit bound Cibacron blue 3GA.[14] In a 6% (w/w) PEG 8000 - 8% (w/w) Dextran T250 system, most of the muscle proteins including LDH partition to the bottom phase, but addition of only 0.05% Eudragit-Cibacron blue (46 mg dye per gram polymer) to the two-phase system resulted in increased LDH partitioning to the top phase ($\Delta K = 2.45$).

Table 2. Purification of recombinant Protein A from cell homogenate by the PEG-Reppal PES 200 two-phase system containing Eudragit bound IgG.

Added Protein A (μg)	Recovered Protein A (μg)			Recovery (%)
	E1	E2	Bottom	
137	80.0	30.1	6.7	80.3
(0.031)[b]	(0.810)[b]	(0.494)[b]		
274	112.0	39.8	53.0	55.3
(0.031)[b]	(0.563)[b]	(0.519)[b]		

Total weight of the system was 5 g (0.025 g of Eudragit bound IgG9.
E1: first fraction of eluate; E2: second fracttion of eluate.
[a] Recovery was calculated from added amount and eluted amount.
[b] Protein A content (mg of Protein A/mg of total protein).
Reproduced with permission from Ref. 13. Copyright 1992, John Wiley & Sons, Inc.

In contrast to the Protein A-IgG system, a single step of affinity extraction with subsequent precipitation of the affinity complex did not result in a very good purification (5 fold) of LDH; this could be due to the lack of absolute specificity of the ligand or the nature of the crude feedstock. In order to obtain a sufficiently pure enzyme, the purification of LDH was carried out in a series of extraction steps.[14]

The muscle extract proteins were first partitioned in a two-phase system containing 6% (w/) PEG, 8% (w/w) dextran and 0.05 M phosphate. The bottom phase with the majority of LDH was contacted with a fresh top phase supplemented with Eudragit-dye. After equilibration, the top phase was separated and `washed´ with a fresh bottom phase. Thereafter, the Eudragit bound affinity complex was recovered from the top phase by precipitation at pH 5. The enzyme was then desorbed from the precipitate by treatment with 0.5 M NaCl with more than 50% recovery. The major loss of LDH activity occurred in the last step due to the denaturation of the enzyme at low pH used for Eudragit precipitation. To avoid this, the protein desorption was also done by the conventional means used in two-phase extraction, i.e. by the addition of salt (11% potassium phosphate) to the PEG phase such that the LDH was transferred to the salt phase. The enzyme was obtained in higher yield (>70%) but with a lower purity.[14] An alternative is to precipitate Eudragit-Cibacron blue at neutral pH by the addition of about 50 mM Ca^{2+} ions and increasing the temperature to 40°C.[17]

CONCLUDING REMARKS

Aqueous two-phase systems offer a number of advantages for downstream processing of proteins, however, they also have their drawbacks. By integrating two-phase extraction steps with other separation methods, a new range of separation techniques may emerge.

ACKNOWLEDGMENT

The authors are grateful to Swedish Agency for Research Cooperation with Developing Countries (SAREC) for financial support.

REFERENCES

1. H.Y. Wang, Integrating biochemical separation and purification steps in fermentation processes. Ann. NY Acad. Sci. 413:313 (1983).
2. B. Mattiasson and O. Holst eds., "Extractive Bioconversion," Marcel Dekker, New York (1991).
3. R. Kaul and B. Mattiasson, Secondary purification, *Bioseparation* 3:1 (1992).
4. H. Hustedt, K.H. Kroner, and M-R. Kula, Applications of phase partitioning in biotechnology, *in*:: "Partitioning in Aqueous Two-Phase System," H. Walter, D.E. Brooks, and D. Fisher, eds., Academic Press, London (1985) pp. 529-587.
5. G. Kopperschläger and G. Birkenmeier, Affinity partitioning and extraction of proteins, *Bioseparation* 1:235 (1990).
6. F. Tjerneld, Aqueous two-phase partitioning on an industrial scale, *in*: "Poly(ethylene glycol) Chemistry: Biotechnical and Biomedical Applications, J.M. Harris, ed., Plenum Press, New York (1992) pp. 85-102.
7. M-R. Kula, Trends and future prospects of aqueous two-phase extraction, *Bioseparation* 1:181 (1990).
8. P. Wikström, S. Flygare, A. Gröndalen, and P-O. Larsson, Magnetic aqueous two-phase separation: a new technique to increase rate of phase separation, using dextran-ferrofluid or larger iron oxide particles, *Anal. Biochem.* 167:331 (1987).
9. S. Flygare, P. Wikström, G. Johansson, and P-O. Larsson, Magnetic aqueous two-phase separation in preparative applications, Enzyme Microb. Technol. 12:95 (1990).

10. P.O. Hedman and J-G. Gustafsson, Protein adsorbents intended for use in aqueous two-phase systems, *Anal. Biochem.* 138:411 (1984).

11. B. Mattiasson and T.G.I. Ling, Efforts to integrate affinity interactions with conventional separation technologies: Affinity partition using biospecific chromatographic particles in aqueous two-phase systems. *J. Chromatogr.* 376:235 (1986).

12. C.A. Ku, J.D. Henry, Jr., and J.B. Blair, Affinity specific protein separations using ligand-coupled particles in aqueous two-phase systems: I. Process concept and enzyme binding studies for pyruvate kinase and alcohol dehydrogenase from Saccharomyces cerevisiae, *Biotechnol. Bioeng.* 33:1081 (1989).

13. M. Kamihira, R. Kaul, and B. Mattiasson, Purification of recombinant protein A by aqueous two-phase extraction integrated with affinity precipitation, *Biotechnol. Bioeng.* 40:1381 (1992).

14. D. Guoqiang, R. Kaul and B. Mattiasson, Integration of aqueous two-phase extraction and affinity precipitation for the purification of lactate dehydrogenase, *J. Chromatogr.* 668:145 (1994).

15. P.A. Harris, G. Karlström, and F. Tjerneld, Enzyme purification using temperature-induced phase formation, *Bioseparation* 2:237 (1991).

16. P.A. Alred, F. Tjerneld, A. Kozlowski, and J.M. Harris, Synthesis of dye conjugates of ethylene oxide-propylene oxide copolymers and application in temperature-induced phase partitioning, *Bioseparation* 2:363 (1992).

17. D. Guoqiang, A. Lali, R. Kaul, and B. Mattiasson, Affinity thermoprecipitation of lactate dehydrogenase and pyruvate kinase from porcine muscle using Eudragit bound Cibacron blue, *J. Biotechnol.* (1994) in press.

MASS TRANSFER IN AQUEOUS TWO-PHASE SYSTEMS

Supriya S. Save and Sanjiv V. Save

Department of Food Science and Technology
University of Reading
P.O. Box 226
Whiteknights
Reading RG6 2AP, United Kingdom

INTRODUCTION

Aqueous two-phase partitioning is being widely used for the purification of proteins. Aqueous two-phase systems offer a gentle environment for cells, cell organelles and biologically active proteins, and partition can be exploited to effect separation otherwise difficult or impossible to achieve. Hence these systems found their early applications in cell biology. Advances in the field of genetic engineering and subsequent developments in enzyme technology have led to investigations of various protein separation processes and their scale up. Presently several areas of aqueous two-phase systems are being investigated:
1. Fundamental analysis of phase separation and protein partitioning
2. Improvement in economy
3. Improvement in the selectivity of extractions
4. Multistage operation

Methods of conventional extraction can be applied to aqueous two-phase systems. Aqueous two-phase systems are characterized by extreme physical properties which differ greatly from those of the conventional liquid-liquid systems. For example, the phases have high viscosities and low density differences. The low interfacial tension in these systems allows the formation of fine droplets even at low power inputs. Fine droplet formation in turn permits high interfacial area, which is desirable for rapid transfer of a solute. The dispersions produced are stable, hence centrifugation is essential for separation. However, such assemblies (consisting of mixers and centrifuges) are expensive for large scale operation. For commercial production of enzymes, often large quantities of fermented broths must be handled, and very high product purity may be needed. For such cases, the mixer-settler apparatus becomes uneconomical, as one set provides only one theoretical stage. In order to purify the desired enzyme from the mixture of enzymes in a fermented broths, more than one theoretical stage is necessary. These requirements can be fulfilled by column type extractors which can be operated in a countercurrent mode. Such extractors have been used for conventional liquid-liquid systems.[1] Preliminary studies on aqueous

Aqueous Biphasic Separations: Biomolecules to Metal Ions
Edited by R.D. Rogers and M.A. Eiteman, Plenum Press, New York, 1995

71

two-phase extraction using conventional contactors such as a Kuhni column, Graesser contactors or mixer-settler batteries have been reported in the literature.[2]   Other conventional contactors such as a spray column, York-Scheibel column, plate column, and packed column to effect aqueous two-phase extraction have also been reported (see Figure 1).[3-6]

For the rational and reliable design of extraction equipment, the knowledge of effective interfacial area (a) and true mass transfer coefficient (k) is highly desirable. Save[7] generated a data base for true mass transfer coefficients of proteins in aqueous two-phase systems in order to study the mass transfer of various proteins across the interface of such systems using a stirred cell (Figure 1).   Since solute diffusivity is an important parameter in mass transfer, the true mass transfer coefficients were measured for different proteins having a wide range of molecular weights. Since physical properties such as viscosity and interfacial tension contribute to the mass transfer coefficients significantly, two phase systems (Poly(ethylene glycol) (PEG)-Dextran and PEG-salt) with a different physical properties were investigated.

The values of mass transfer coefficient depend upon the flow field in the vicinity of the interface. In spite of the incomplete understanding of such flow fields, substantial progress has been made in modelling the process of mass transfer, and a number of theories are reported in the published literature. In each of these theories, the flow structure in the vicinity of the interface is assumed and thereupon the theory is developed. In this way a variety of theories have been proposed. One of the models is likely to be valid under a particular set of conditions, and the selection of model depends strongly on the flow structure. A quantitative understanding of the scales and velocities of eddies responsible for mass transfer is necessary. Joshi[8] attempted to understand the flow field in the vicinity of the interface, to confirm the validity of the existing mass transfer theories.

In this article, several theories of mass transfer as they are applied to aqueous two-phase extraction are discussed followed by evaluation of the performance of various liquid-liquid contactors using these systems.

EVALUATION OF THEORIES OF MASS TRANSFER FOR PROTEIN MASS TRANSFER

Film theory

In the film theory, a thin film in the vicinity of the interface is assumed to provide the entire resistance to mass transfer. This film is assumed to be completely stagnant, and mass transfer occurs by molecular diffusion and under steady-state conditions. The liquid side mass transfer coefficient is given by the equation:

$$k = D/\delta \qquad (1)$$

A comparison of the mass transfer coefficients predicted by the film theory with the experimental mass transfer coefficients suggests that the predicted values are substantially lower (by a factor of even one thousand) than the experimental values. This observation indicates that the role of eddy diffusion is very important in understanding the protein mass transfer.[8]

Surface renewal model

Due to its oversimplification, the film model does not find any experimental or theoretical support. Danckwerts[9] introduced the random surface-renewal concept to describe

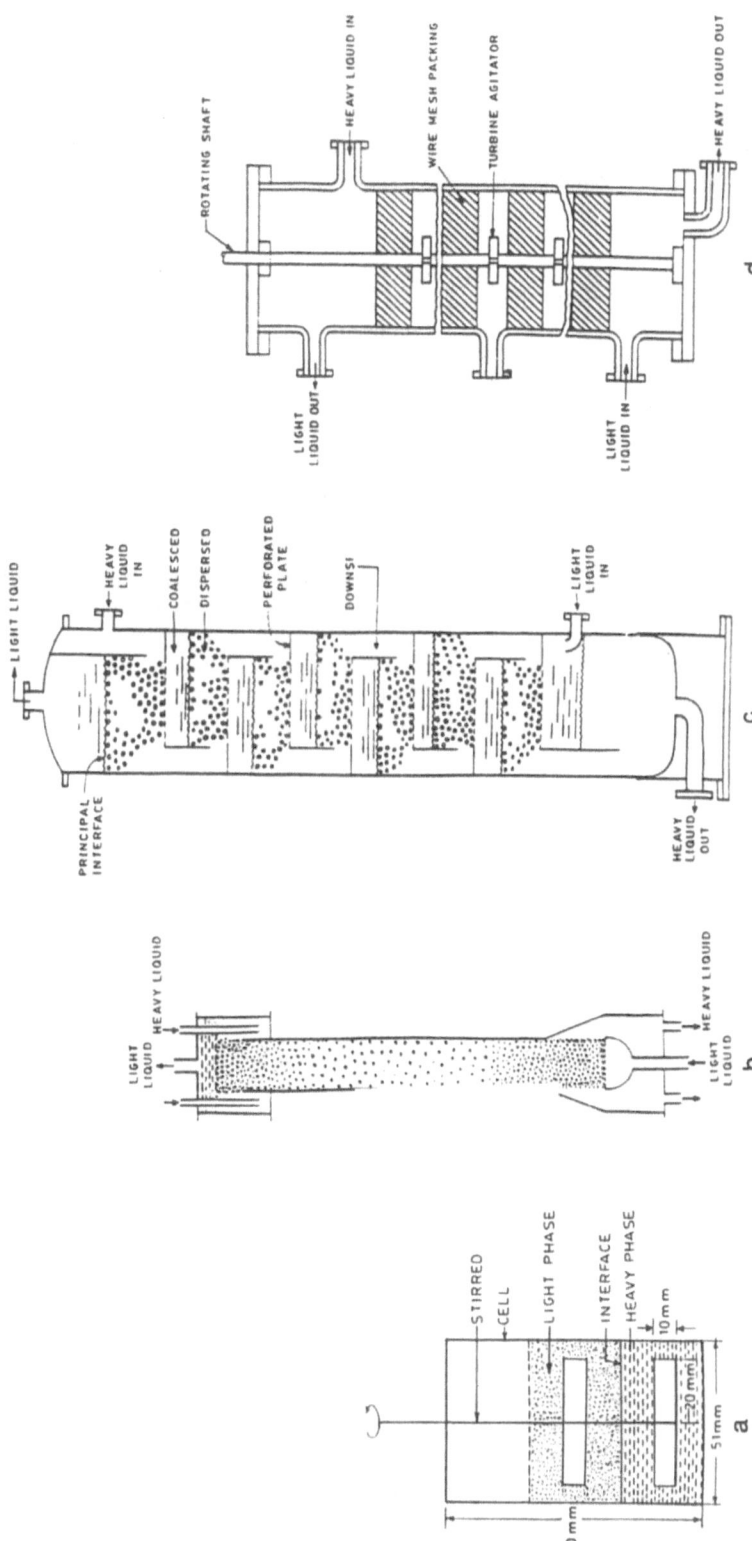

Figure 1. Various liquid-liquid contactors: a) stirred cell, b) spray column, c) plate column, and d) York-Scheibel column.[4] (Reprinted by permission of Kluwer Academic Publishers)

the mechanism of gas absorption, where mass transfer occurs from the gas-liquid interface to the bulk of the liquid. The basis for this random surface renewal concept was that the liquid elements at the gas-liquid interface are exchanged randomly in time with elements from the bulk of the turbulent liquid. For the relatively short residence times of these liquid elements at the gas-liquid interface, the elements were assumed to be stagnant or plug flowing, and the mass transfer was calculated using the penetration. Here, a definite pattern of age distribution was assumed for these elements given by

$$\phi = s \, e^{-st} \tag{2}$$

The predicted values from this model for protein mass transfer coefficients in all phase systems considered were 2 to 10 times higher than the experimental values.[8]

Using the data of Save[7] for PEG-Dextran systems (shown in Table 1) using the solutes cytochrome-c, α-amyloglucosidase, and β-galactosidase, Joshi *et al.* further applied mass transfer theories which are proposed for gas-liquid and liquid-liquid systems to understand their applicability towards protein mass transfer in aqueous two-phase systems.[8] In addition, validity of these theories for PEG-salt systems (Table 1) was also simultaneously ascertained. The scope of this chapter is not to describe each theory in detail and its applicability for predicting protein mass transfer in aqueous two-phase systems. Therefore, the predictions of Joshi et al.[8] using the following theories are briefly described, mainly to highlight the limitations of each theory to predict true protein mass transfer coefficient in aqueous two-phase systems: (i) Small eddy model of Lamont and Scott,[10] (ii) Large eddy models of Fortescue and Pearson[11] and Luk and Lee,[12] and (iii) the Levich[13] and Levich-Davies models.[14] All these models attempt to relate the rate of surface renewal to the flow parameters in the vicinity of interface.

Small eddy model

Predicted values of protein mass transfer for systems of low viscosity (I and II), matched with the experimental data very well as compared to high viscosity systems (III and IV), for which predictions were very poor. In the case of a low molecular weight solute, cytochrome-c, the predictions were lower than the experimental values while in the case of high molecular weight proteins, α-amyloglucosidase and β-galactosidase, they were higher than the experimental values. When the average power consumption was used for predictions, the calculated mass transfer values were always higher than those predicted using local power consumption.[8]

Large eddy model

The predicted values for protein mass transfer coefficient with the large eddy model in all systems considered were 3 to 30 times higher than the respective experimental values.[8]

Levich model

The observed difference between the predicted and experimental values was large for systems I and II, and the agreement was fairly good for systems III and IV. The film theory did not consider the role of eddy velocity towards mass transfer. The Levich model, on the other hand, predicts an overdependence on the eddy velocity ($k \propto u'^2$). The prediction of both the models showed disparity with the experimental data. To understand this disparity, the values of u' were experimentally measured and used directly to check the

Table 1. Physical properties of phase systems used for mass transfer studies.[7]

PEG 4000-dextran phase system

System	Viscosity (mPa·s)		Density (kg/m³)		Interfacial Area (mN/m)
	PEG rich phase	Dextran rich phase	PEG rich phase	Dextran rich phase	
I	5.01	28	1028	1058	0.03×10^{-2}
II	5.35	60	1029	1060	0.81×10^{-2}
III	5.68	155	1030	1090	2.00×10^{-2}
IV	6.30	263	1039	1105	4.10×10^{-2}

PEG 4000-sodium sulfate phase system

System	Viscosity (mPa·s)		Density (kg/m³)		Interfacial Area (mN/m)
	PEG rich phase	Dextran rich phase	PEG rich phase	Dextran rich phase	
I	12.2	1.39	1074	1123	0.0059
II	20.3	1.47	1075	1126	0.1500
III	24.1	1.49	1079	1140	0.2400

dependence of k on u'. The following correlation was obtained for PEG-Dextran system:[8]

$$k = 0.5 (D/v)^{0.5} u' \qquad (3)$$

In this case, the parity was reasonable over the entire range of mass transfer coefficient. The exponent of one for u' is between the exponents predicted by surface renewal theories and the Levich theory. Further, the mass transfer coefficient is proportional to the square root of the diffusivity.

Similar analyses for mass transfer in PEG-sodium sulfate systems suggested that the behaviour is similar to that of PEG-Dextran system, yielding the following equation:

$$k = 0.2 (D/v)^{0.5} u' \qquad (4)$$

According to Equations 3 and 4, the actual velocity with which an eddy approaches the interface plays a significant role in the mass transfer.[8] In summary, the results of Joshi et al.[8] showed that the dependence of mass transfer coefficient on the square root of the diffusivity also holds for macromolecules like proteins in these aqueous liquid-liquid systems having very high viscosities. The mass transfer coefficient was also found to be proportional to the eddy velocity in the direction normal to the interface (k ∝ u').

LIQUID - LIQUID CONTACTORS FOR AQUEOUS TWO-PHASE EXTRACTION

The behaviour of various conventional liquid-liquid contactors has been systematically investigated for aqueous two-phase extraction of proteins as noted in the introduction. The variables investigated can be classified as:

1.   Engineering variables such as dispersed phase velocity, column height, sparger design, scale-up, *etc.*
2.   Two-phase system variables such as components of system, compositions, protein molecular weights, *etc.*

These previous investigations suggest that liquid-liquid contactors investigated so far perform in similar fashion though the achievable rates of mass transfer vary. In this section, the specific effect of the above mentioned variables on mass transfer coefficient is discussed, in addition to a summary of various correlations proposed for estimating dispersed phase hold-ups and mass transfer coefficients in the studied contactors.

In all previous investigations, liquid-liquid contactors have been operated in semi-batch mode with the PEG-rich upper phase as the dispersed phase.

Effect of dispersed phase velocity, V_d

Figures 2 and 3 show the typical variation of dispersed phase hold-up (ε_D) and overall protein mass transfer coefficient ($K_D a$) for various contactors and for various aqueous two-phase systems. The dependence is almost linear. Kumar and Hartland[15] indicated that the drop size, d_p, decreases with an increase in V_d. Hence, an increase in V_d increases the number of drops, resulting in a proportional increase in the value of ε_D. Since the value of a is given by

$$a = \varepsilon_D / 6 d_p \qquad (5)$$

a also increases resulting in proportionate increase in $K_D a$.[16]

Effect of column height

Conventional differential liquid-liquid contactors require certain minimum height in order to achieve completely developed continuous and dispersed phase flow patterns. Underdeveloped flow patterns at lower column heights result in very poor mass transfer rates. Reported data for various aqueous two-phase extractions in spray column suggest that a column height of 500 mm is sufficient. This minimum height arises from the observation that fully developed flow patterns are achievable in such contactors when operated with column height 5-10 times the column diameter.[17] Hence, a column height of 10 times the column diameter could be considered an appropriate criterion.

Effect of sparger design

The effect of number of holes and its diameter has been widely investigated for aqueous two-phase extactions using spray column and aphrons[5,18,19] and suggests that ε_D and $K_D a$ increases with an increase in number of holes. This behaviour has been explained by Jafarabad, *et al.*[5] and Pawar, *et al.*[19] as follows:

In the case of single hole sparger, one jet is formed which disintegrates into drops at the end of the jet. The drops rise upwards with some displacement in the radial direction. Therefore, ε_D increases with an increase in the column height up to a certain length. Beyond this height, the uniformity in the hold-up is achieved and the change with respect to column height becomes negligible. Thus, the column height can be divided into

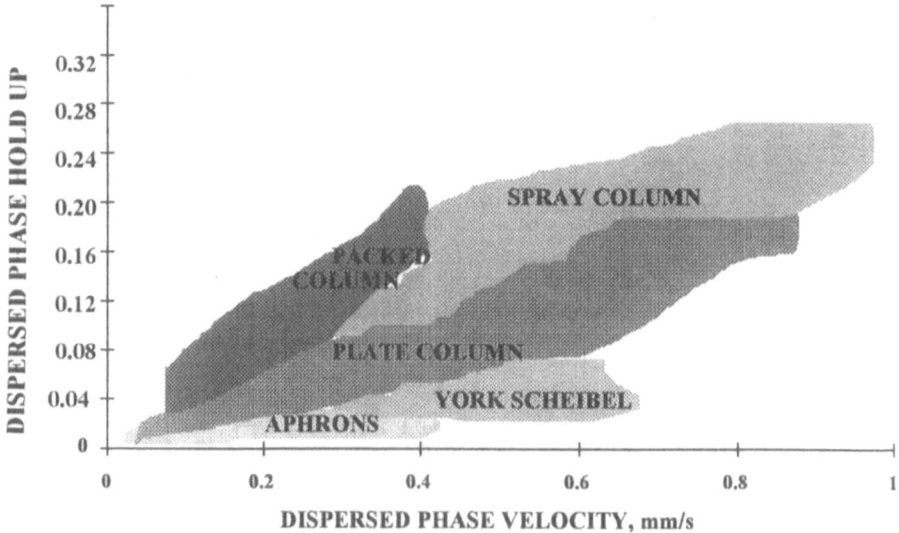

Figure 2. Typical variation of dispersed phase hold-up with dispersed phase velocity in various liquid-liquid contactors.[3-6,8]

two regions: the sparger region and the bulk region. In the sparger region, the development of hold-up occurs while in the bulk region, hold-up remains independent of the column height. For the case of single hole sparger, large volume of continuous phase remains unutilised at the bottom (sparger region) owing to jetting of the dispersed phase. On the other hand, for multiorifice spargers, the height of the sparger region is lower in comparison because dispersed phase enters the column more uniformly. Hence, ε_D increased with the increase in number of holes.

The increase in the value of $K_D a$ with an increase in the number of holes has been attributed to the increase in the ε_D.

Effect of scale up

Investigations reported for scale up of spray and packed columns suggest that $K_D a$ and ε_D remain unaffected by column diameter, obviously due to the complete development of flow patterns.

The foregoing discussion suggests that contactors operated with multiorifice sparger at high superficial dispersed phase velocities will result in maximum mass transfer rates. Despite the fact that true mass transfer could be predicted from the model proposed by Joshi et al.[8], the design of the contactor is still difficult. A procedure to aid the design of liquid-liquid contactor is first to estimate ε_D and k then to estimate a.

It is very difficult to estimate ε_D and a (which in fact depends on the ε_D). Though a wide range of empirical and semiempirical correlations are available for these estimations, their applicability is restricted by the diversity of two-phase systems encountered in bioprocessing and their properties. In order to achieve accurate estimation of ε_D, more knowledge of hydrodynamic behaviour is needed. Joshi et al.[4] have suggested

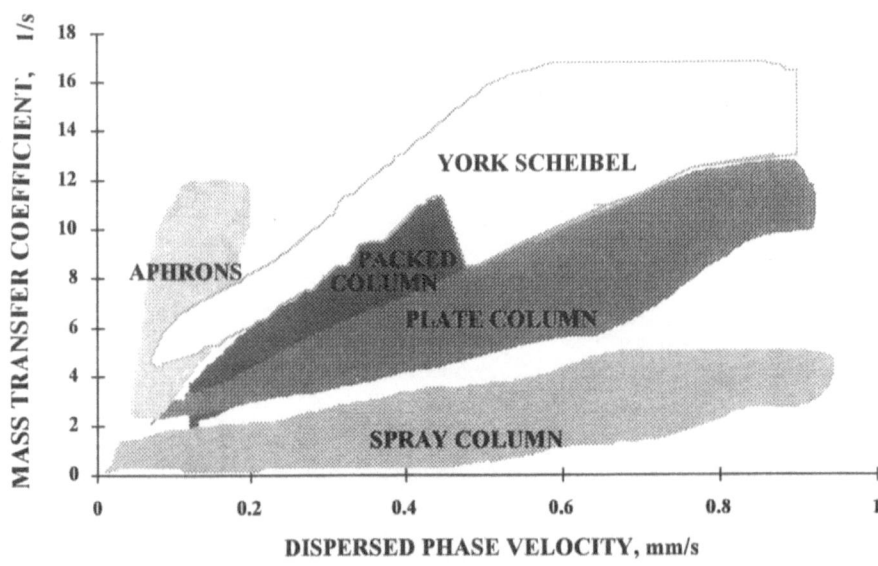

Figure 3. Typical variation of dispersed mass transfer coefficient with dispersed phase velocity in various liquid-liquid contactors.[3-6,8]

investigations in the following areas which will be useful for more reliable design of liquid-liquid contactors for aqueous two-phase extraction:

1. Drop dynamics in aqueous two-phase systems should be thoroughly investigated. Such studies will allow the prediction of minimum and critical jetting velocity, minimum jet length and drop formation caused by jet break up. Also, methodology for predicting drop size and terminal drop velocities should be established. Moreover, understanding of mass transfer during drop formation will help in developing rational design equations.

2. Flooding characteristics of these contactors and coalescence of drops needs to be investigated from the point of view of achieving higher operating range. Some aspects of drop dynamics are investigated by Bhavsar[20] while Pawar et al.[19] employed enlarged exit section to increase flooding limits.

3. For successful implementations of these two above-mentioned investigations, one needs reliable data of physical properties of various phase systems correlated to their respective composition.

Effect of phase compositions

In case of PEG-salt systems, an increase in tie line length leads to an increase in dispersed phase viscosity, interfacial tension as well as density difference. The net result of these factors is an increase in the average drop size and hence drop velocity.[20] These effects lead to lower ε_D and hence lower a, and thus lower $K_D a$, as explained above.

In addition to this effect, a change in phase composition will affect overall mass transfer coefficient in another way. The overall mass transfer coefficients given by

$$\frac{1}{K_D} = \frac{m}{k_c} + \frac{1}{k_D} \qquad (6)$$

The value of partition coefficients of the proteins in the system shown in Table 1 were on the order of 10^{-4}. Thus, m/k_c is very low thus effectively making $m/k_c << 1/k_D$, implying that the mass transfer process is controlled by the resistance in the dispersed phase. Thus, change in μ_d causes changes in i) diffusivity of the protein in the dispersed phase. An estimate of protein diffusivity is:[22]

$$D = \frac{9.4 \times 10^{-15} T}{\mu_d \, M^{0.33}} \qquad (7)$$

which suggests that diffusivity is inversely proportional to the dispersed phase viscosity; ii) interfacial area, a as explained above; and iii) a possible decrease in the true mass transfer coefficient, k_D.

For the case of polymer-polymer systems, continuous phase viscosity changes substantially with change in the tie line length in contrast to polymer-salt systems. Since drop velocity decreases with an increase in continuous phase viscosity, hold-up values do not change as monotonically as in the case of polymer-salt systems.

Intensification of Mass Transfer

Recently Save et al.[18] developed a technique to intensify the mass transfer in aqueous two-phase extraction. This technique involved conversion of the dispersed phase into colloidal gas aphrons. Colloidal gas aphrons consist of individual aphrons composed of gas (air) entrapped in a soapy film stabilized by surfactant. The properties of colloidal gas aphrons have been reported[23,24] and these properties offer the following advantages:
1.   Lower disengagement time owing to higher phase density difference
2.   Higher phase purity
3.   Considerably higher interfacial area compared to conventional contactors
4.   Higher utilization of costly polymers due to enhanced mass transfer rate, thereby increasing the economic viability of aqueous two-phase extraction.
This scheme yielded mass transfer coefficients 20 times higher than those obtained using a spray column.

CONCLUSION AND THE SCOPE FOR FUTURE WORK

Although behaviour of various liquid-liquid contactors for the extraction of proteins using aqueous two-phase systems is very well understood, a priori design is still a distant goal. Inspite of the knowledge of the protein mass transfer across the interface in aqueous two-phase systems, lack of understanding in drop dynamics has prevented the formation of a reliable set of design equations. Additionally, such efforts have to be supplemented by developing a model for continuous aqueous two-phase extrations.[25]

Another major hurdle in the applicability of various design equations reported in the literature is the determination of the diffusivity of protein molecules. The equation given by Geankoplis[22] is commonly used by researchers, although its validity is questionable. Although this equation has a viscosity term, it implies the viscosity of the solvent and not the solution (e.g. a PEG solution). Since PEG concentrations are substantially high, the orientation of PEG molecules in aqueous solutions may be sufficiently complex to affect the diffusion of large protein molecules. Moreover, behaviour of proteins, such as their orientation, tendency to form polymers, *etc.* in PEG solution is still unknown. Other equations reported in the literature predict protein diffusivity within 5 to 10 % of that predicted by Equation 7.[7]   A methodology is needed for determining apparent diffusivity of proteins specifically in aqueous two-phase systems.

NOMENCLATURE

a	-	Effective interfacial area, m^2/m^3
D	-	Diffusivity of protein, m^2/s
d_p	-	Drop diameter, m
K	-	Overall mass transfer coefficient, m/s
k	-	True mass transfer coefficient, 1/s
k_c	-	Continuous phase mass transfer coefficient, 1/s
k_d	-	Dispersed mass transfer coefficient, 1/s
m	-	Partition coefficient of protein
M	-	Molecular weight of protein
s	-	Surface renewal rate, s
t	-	Exposure time of an eddy, s
T	-	Temperature, K
u'	-	Rms velocity at a distance y from the interface, m/s
V_d	-	Superficial dispersed phase velocity, m/s

Greek

ε_D	-	Fractional dispersed phase hold-up.
ϕ	-	Age distribution
Δ	-	Film thickness, m
μ	-	Viscosity of the lighter phase, mPa.s
ν	-	Kinematic viscosity, m/s
ρ	-	Density of the phase, kg/m^3
σ	-	Interfacial tension, mN/m

REFERENCES

1. R. E. Treybal, "Liquid-Liquid Extraction," McGraw Hill, New York (1963).
2. J. D. R. Coimbra, J. Tommes, and M.-R. Kula, Continuous separation of whey proteins with aqueous two-phase systems in a Graesser contactor, *J. Chromatogr. A* 668:85 (1994).
3. S. B. Sawant, J. B. Joshi, and S. K. Sikdar, Hydrodynamics and mass transfer characteristics of spray columns for two phase aqueous extraction, *Biotechnol. Bioeng.*, 36:109 (1990).
4. J. B. Joshi, S. B. Sawant, K. S. M. S. Raghav Rao, T. A. Patil, K. M. Rostami, and S. K. Sikdar, Continuous counter-current two-phase aqueous extraction, *Bioseparation*, 1:311 (1990).
5. K. R. Jafarabad, S. B. Sawant, and J. B. Joshi, Design and scale-up of spray extraction columns for aqueous two phase extraction, *Chem. Eng. Sci.*, 47:57 (1992).
6. K. R. Jafarabad, T. A. Patil, S. B. Sawant S.B., and J. B. Joshi, Enzyme and protein mass transfer in York-Scheibel columns, *Chem.Eng.Sci*, 47:69 (1992).
7. S. S. Save, Ph.D. Thesis, University of Bombay (1994).
8. J. B. Joshi, S. S. Save, R. B. Desai, and S. B. Sawant, True mass transfer coefficients

for proteins in aqueous two phase extraction: Theories of mass transfer revisited, *Chem. Eng. Sci.*, Under review.

9. P. V. Danckwerts, Significance of liquid film coefficients in gas absorption, *Ind. Eng. Chem.*, 43:1460 (1951).

10. J. C. Lamont, and D. S. Scott, An eddy cell model of mass transfer in the surface of a turbulent liquid, *AIChE J.*, 16:513 (1970).

11. G. E. Fortescue, and J. R. A. Pearson, On gas absorption into a turbulent liquid, *Chem. Eng. Sci.*, 22:1163 (1967).

12. S. Luk, and Y. H. Lee, Mass transfer in eddies close to air-water interface, *AICHE J.*, 32:1546 (1986).

13. V. G. Levich, "Physicochemical Hydrodynamics," N. J. Tiffs, ed., Prentice-Hall, Englewood (1962).

14. J. T. Davies, Turbulence phenomena at free surfaces, *AIChE J.*, 18:169 (1972).

15. R. Kumar, and S. Hartland, Prediction of drop size produced by multiorifice distributor, *Trans. Inst. Chem. Engr.* 24:213 (1982).

16. K. S. M. S. Rao, D. S. Szlag, S. K. Sikdar, J. B. Joshi, and S. B. Sawant, Protein extraction in a spray column using a polyethylene glycol-maltodextrin system, *Chem. Eng. J.* 46:B75 (1991).

17. L. K. Doraiswamy, and M. M. Sharma, "Heterogeneous Reactions: Analysis, Examples and Reactor Design, Vol. 2: Fluid-Fluid-Solid Reactions," Wiley, New York (1984).

18. S. V. Save, V. G. Pangarkar, and S. Vasantkumar, Intensification of mass transfer in aqueous two phase systems, *Biotechnol. Bioeng.*, 41:72(1993).

19. P. A. Pawar, K. R. Jafarabad, S. B. Sawant, and J. B. Joshi, Enzyme mass transfer coefficient in aqueous two phase systems: Spray extraction columns, *Chem. Eng. Commun.*, 122:151 (1993).

20. P. A. Bhavsar, P.A., Ph. D. (Tech.) Thesis, University of Bombay (1993).

21. R. Clift, J. R. Grace, and W. E. Weber, "Bubbles, Drops and Particles," Academic Press, New York (1978).

22. C. J. Geankoplis, Transport Processes: Momentum, Heat and Mass Transfer, Allyn and Bacon, London (1983).

23. S. V. Save, and V. G. Pangarkar, Characterization of colloidal gas aphrons, *Chem. Eng. Commun.* 127:35 (1994).

24. S. V. Save, V. G. Pangarkar, and S. Vasnatkumar, Liquid-liquid extraction using aphrons, *Separ. Tech.* 4:104 (1994).

25. J. A. Asenjo, E. Leser, and B. Andrews,, New perspective in bioseparations, *in:* "Separation Technology: Next Ten Years," J. Garside, ed., I.Chem.E., United Kingdom (1994).

METAL EXTRACTION IN TWO-PHASE WATER-POLY(ETHYLENE GLYCOL)-SALT SYSTEMS

Boris Ya. Spivakov, Tatjana I. Nifant'eva and Valery M. Shkinev

Vernadsky Institute of Geochemistry & Analytical Chemistry
Academy of Sciences
Moscow V-334, Kosygin str. 19, 117075
Russia

INTRODUCTION

Liquid-liquid extraction, one of the most effective technological and analytical techniques for the separation of substances, is based in most cases on their distribution between an aqueous solution and a water-immiscible organic solvent. The distribution of a substance between an aqueous phase and an organic one significantly depends on the difference between its hydration and solvation energies in these phases. The conventional method to achieve extraction is to adjust the two parameters by the addition of reagents, which form compounds with the substance to be extracted having low hydration and high solvation energies. However, application of systems involving an aqueous and an organic solution does not always result in desirable separations, especially if it is difficult or impossible to form compounds possessing the properties mentioned. This takes place, for example, when it is necessary to extract highly charged inorganic complex ions or water-soluble organic substances and their complexes with metal ions. Such reagents could extend the sphere of application of liquid-liquid extraction to separation and concentration of inorganic ions. Separation in this case can be based on the usage of heterogeneous liquid systems in which both phases contain significant amounts of water to decrease the effect of hydration on the transfer of complexes with water-soluble reagents from one phase to another.

Polymer-based two-phase aqueous systems are widely applied to the gentle partition of biopolymers, cells, products of biotechnology, etc.[1,2] Application of such systems for metal separation was suggested by the authors in papers[3-5] where unequal distribution of metal ions and their complexes with analytical organic reagents and inorganic ions was shown to occur between two phases formed due to the phase separation of aqueous solutions of poly(ethylene glycol) (PEG) in the presence of some inorganic salts or another polymer.

It is also of interest for technological processes to use liquid-liquid extraction systems without organic solvents which are often volatile, flammable, or explosive. From an economical point of view, PEG is attractive because of its low cost, and if the polymer itself serves as extractant there is no need to add any other organic reagents.

Aqueous Biphasic Separations: Biomolecules to Metal Ions
Edited by R.D. Rogers and M.A. Eiteman, Plenum Press, New York, 1995

This paper describes the extraction of some metals in PEG-inorganic salt-water systems. The possibility of practical applications of such systems is discussed.

RESULTS AND DISCUSSION

Phase Separation

Extraction systems based on various inorganic salts could be of practical interest. From the data available in the literature it was not always clear which salts and which salt concentrations could result in salting out of PEG. That is why one of the tasks of this work was to investigate phase equilibria in water-salt solutions of PEG in order to choose the conditions for obtaining heterogeneous systems.

Binodal curves, demarcating the regions of equilibria of two liquid phases and those of homogeneous solutions for systems containing PEG-2000 and K^+, Na^+, NH_4^+ salts of different acids, are presented in Figure 1. The salting-out ability of anions was evaluated according to the values of minimal molality of salt solution (C_{min}) necessary for phase separation in the systems containing 15% polymer:

Anion	CO_3^{2-}	HPO_4^{2-}	SO_4^{2-}	F^-	ClO_4^-	SCN^-
C_{min}	0.69	0.72	1.03	3.36	5.72	9.06

The obtained order: $CO_3^{2-} > HPO_4^{2-} > SO_4^{2-} > F^- > ClO_4^- > SCN^-$ coincides with the well-known Hofmeister series for biopolymers.[6] It should be noted, however, that it is impossible to make a conclusion about the salting-out ability of the salt on the basis of the anion position in the Hofmeister series. Indeed, we could not obtain heterogeneous systems for nitrates, chlorides and bromides, though they occupy an intermediate position between F^- and SCN^-, for which two-phase systems can be obtained.

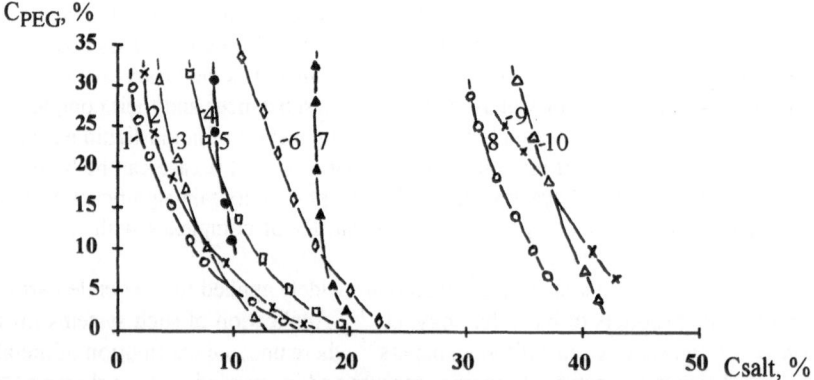

Figure 1. Phase diagrams for PEG-2000-salt-acid (base)-water systems. 1: $(NH_4)_2SO_4$, 1 M NaOH; 2: $(NH_4)_2HPO_4$; 3: K_2CO_3; 4: $(NH_4)_2SO_4$; 5: NH_4F; 6: $(NH_4)_2SO_4$, 1 M H_2SO_4; 7: KSCN, 0.6 M HCl; 8: $NaClO_4$; 9: KI, 0.6 M HCl; 10: KSCN.

Phase equilibria in systems comprised of PEG and sulfate, thiocyanate, iodide, or fluoride solutions in the presence of significant amounts of acids or bases were investigated.

The salting-out effect of OH⁻ ion, whose position in the Hofmeister series is before F⁻, causes increased heterogeneous regions of the systems based on $(NH_4)_2SO_4$ in the presence of NaOH (Figure 1, Curves 1 and 4). In the presence of an acid, a decrease in the heterogeneous region of the sulfate-based system is observed (Figure 1, Curves 4 and 6). However, for the systems comprised of thiocyanate or iodide the increasing acid concentration results in a strong salting-out effect for PEG (Figure 1, Curves 7, 9, and 10). Moreover, systems based on PEG-2000 and KI cannot be obtained at all in the absence of acid.

The possibility of obtaining two-phase systems with the salts of different metals was investigated for sulfate solutions. The salting-out ability order was obtained by the method used for anions:

Cation	Zn^{2+}	Co^{2+}	Ni^{2+}	Mg^{2+}	Cu^{2+}	Cd^{2+}	NH_4^+	Li^+
C_{min}	0.40	0.62	0.62	0.70	0.71	0.75	1.03	1.13

The salting-out ability of cations is less clear than in the case of anions and does not significantly affect the width of the heterogeneous region. Thus, considering the possibility of obtaining two-phase systems with the salt of some metals, not the salting-out ability of the given cation, but the solubility of the salt, should be taken into account. For example, due to a relatively low solubility of K_2SO_4 and Na_2SO_4 heterogeneous systems with these salts were not obtained.

In order to estimate the applicability of the studied aqueous PEG-salt systems on a process scale, systems settling under gravity have been examined. The phase separation times were measured for PEG-$(NH_4)_2SO_4$ systems in the presence and absence of sulfuric acid. The components were stirred in a static mixer and the emulsion was instantly poured into a measuring cylinder for the phase separation studies. Comparative studies were made to evaluate the separation times for undiluted tributyl phosphate (TBP)-aqueous $(NH_4)_2SO_4$ solution and an undiluted petroleum sulfoxide (PS)-aqueous $(NH_4)_2SO_4$ solution system. The dispersed phase separation velocities for the systems containing 10% ammonium sulfate and 40% aqueous PEG, TBP, or PS were found to be 1.39, 2.85, and 0.79 m/h, respectively. The dispersed phase separation velocity for the 40% PEG-40% K_2CO_3 system is 1.15 m/h. The calculations of the dispersion numbers by Rogers and coworkers[7] have shown that their values are comparable to many oil-water systems in use today. It follows from the above-mentioned results and from the phase densities measured that the PEG-based systems can be employed on a process scale using conventional extraction equipment. Studies of a PEG-2000-$(NH_4)_2SO_4$ system containing NH_4F made with mixer-settler contractors have shown that the system behavior is similar to that of a TBP-based system with the same salt phase. Specially designed extraction plants are used for the recovery and purification of biochemical substances in such systems.[1]

Metal Extraction in PEG-Based Systems in the Absence of Extractant

For selective extraction of metals of interest, conditions should be provided not only for the effective extraction of their complexes with a certain extractant added but also for low distribution coefficients (D) of metals which do not form such complexes, i.e., for low-blank extraction. Detailed studies of the blank extraction of various metals for the system PEG-2000-$(NH_4)_2SO_4$-H_2O in the pH range from 2 to 6 have shown that the D values depend on many factors: phase composition (in particular PEG concentration in the polymer-rich phase and $(NH_4)_2SO_4$ distribution between the phases, D_{SO4}), pH, and metal concentration. The data presented in Figure 2 illustrate such complex dependencies. The equations have been proposed[8] which enable the selection of conditions for minimal blank extraction. For example, it follows from one of the equations and from experimental data that at trace metal concentrations and pH less than 4, or at relatively high metal concentrations ($> 10^{-4}$ M) and any pH, the D values are proportional to the D_{SO_4} value, the coefficient of proportionality being 0.44. Thus, moving from the binodal curve into the heterogeneity region causes a decrease of the blank extraction.

A study of the PEG-(Na-sulfate dextran)-NaCl-H_2O system[9] has shown that a different pH dependence is observed in a system based on two polymers. The blank extraction decreases at pH > 4 which may be important for some separations.

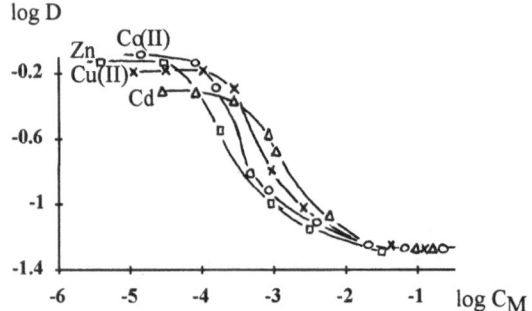

Figure 2. Dependence of distribution coefficients of metals on their total concentration in the system PEG-$(NH_4)_2SO_4$-H_2O.

Extraction of Metals with Water-Soluble Organic Reagents

The distribution of a wide range of organic water-soluble reagents was investigated. All reagents in this study containing aromatic rings, irrespective of the number of polar or dissociating groups, are extracted in the polymer phase with distribution coefficients more than 100. On the other hand, the D values for aliphatic complexing agents are less than 1. It seems that for these compounds in which the nonpolar part of the molecule is shielded by acetate groups and not accessible for hydrophobic interactions with PEG, the factor which determines their distribution is the difference between the hydration ability of the two phases. In the case of the aromatic reagents, the energy of hydrophobic interaction between -CH_2CH_2- groups in the PEG chains and bulky nonpolar groups of the compound being transferred to the polymer phase completely compensates for small unfavorable changes in the hydration of the hydrophilic groups. Thus, one of the conditions responsible for the transfer of an organic compound into the polymer phase is the presence of a nonpolar part of the molecule, which is accessible for hydrophobic interaction. The number of polar or dissociating groups in the molecule does not significantly affect the extractability of the compound.

For metal extraction in aqueous two-phase systems, we suggested[3-5] application of a wide range of aromatic chelating dyes, which had been effectively used as photometric analytical reagents due to their high metal complexation constants and selectivity. Among them are Alizarin Complexone, triphenylmethane dyes (Methylthymol Blue, Xylenol Orange, etc.), arsenazo group reagents (Thoron, Arsenazo M, Arsenazo III, etc.), *o*-hydroxyphenylimino-*N*,*N*-diacetic acid, and other photometric reagents. The data on the use of these reagents for extraction of actinides, lanthanides, and transition metals from sulfate, carbonate, phosphate, and other salt solutions have been reviewed in papers.[10,11]

The quantitative extraction of metal complexes with water-soluble organic reagents demonstrates some advantages of the two-phase aqueous systems. Indeed, in traditional water-organic solvent systems, the extraction of such complexes is often difficult. The metal-to-reagent ratio in the complexes is usually 1:1, so the compounds formed by tri- or tetra-charged cations have a positive charge, and the presence of hydrophobic counter-anions (ClO_4^-, CCl_3COO^-, etc.) is necessary. For compensation of sulfonic group charge, hydrophobic cations must also be added (such as diphenylguanidine). Furthermore, the solvent used must have substantial solvating ability. Even if all of these conditions have been met, high distribution coefficients are still not always achieved.

Metal Extraction in the Presence of Halide and Thiocyanate Ions

Extraction of halide and thiocyanate complexes of metals is widely used for their separation. The most important are complex metal-halide acids which are extracted only with the use of donor-active extractants protonated in acidic media. Singly charged metal-containing anions are easily extracted in the form of complex acids by any oxygen-bearing extractant, whereas only the most effective extractants (ketones, organophosphorus compounds) should be used to extract dicharged anionic complexes. Although PEG is a weakly basic polyether, we expected that in biphasic aqueous systems it should effectively extract both mono- and dicharged anionic complexes of various metals. Indeed, Cu(II), Co(II), Zn, Fe(III), Mo(V), and In, which form such anions in thiocyanate solutions, are extracted in a PEG-$(NH_4)_2SO_4$ system containing 1 M NH_4SCN and 1-2 M H_2SO_4 with distribution coefficients higher than 100. It was shown that D values for the dicharged thiocyanate complex of copper(II) are 3 to 4 orders of magnitude higher in PEG solution than in monomeric diisopropyl ether, and even higher than for such an effective water-insoluble extractant as TBP.

Figure 3 illustrates the extraction of various metal ions as a function of HCl concentration in the PEG-KSCN-H_2O system. It was of interest that in this system inorganic salt simultaneously supplies the anion for extraction of a metal complex and salts out PEG. It is seen that Zr, Hf, Sc, and lanthanides are extracted into the PEG-phase in such a system, while in the $(NH_4)_2SO_4$-based system they are not extracted due to their strong complexation by sulfate in the salt phase of the system.

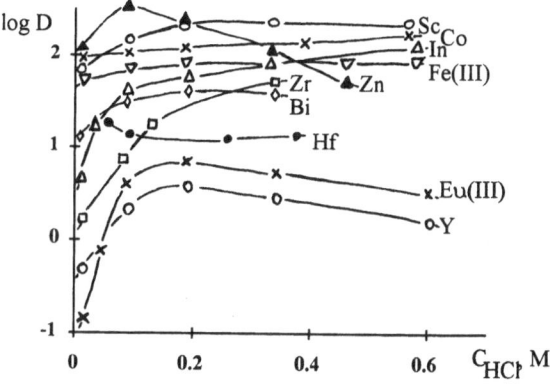

Figure 3. Metal extraction in the PEG-KSCN-H_2O system.

The distribution coefficients in halide-containing solutions increase in the order $Cl^- < Br^- < I^-$. The quantitative extraction of Tl(III) and Bi from 0.1-2.5 M NH_4I and 1.0-2.7 M H_2SO_4 was observed.[12] From 2.5 M NH_4I and 2.7 M H_2SO_4 solutions, In and Sb(III) are extracted with D > 100, Cu(I), Cd, and Zn with D > 10.

Metal Recovery and Purification

Two variants of processes based on utilization of two-phase aqueous systems can be applied: extraction of a desirable macro component into the PEG phase and extraction of impurity components of the salt solution containing the macro component to be purified. The loading capacity of the polymer phase is an important characteristic of the PEG phase for the first variant.

Extraction of niobium from sulfate-fluoride solutions has been studied as an example of metal purification by its extraction into the polymer phase. Extraction has been performed from aqueous $(NH_4)_2SO_4$ solutions at a molar ratio of Nb:F = 1:6 in the presence and absence of H_2SO_4 by 40% PEG solution (Figure 4).

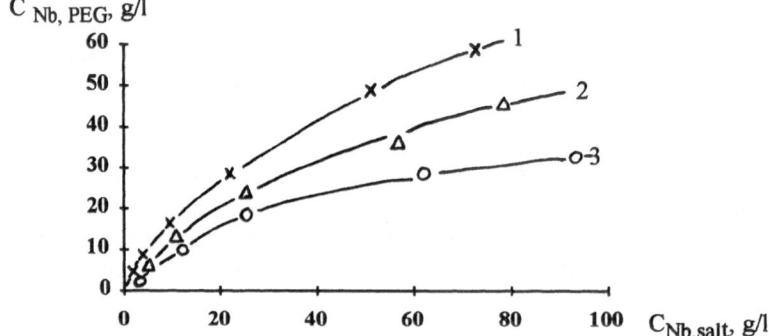

Figure 4. Extraction of niobium by 40% PEG solution from sulfate-fluoride solutions with a molar ratio of Nb:F = 1:6. $(NH_4)_2SO_4$ concentration in the salt phase = 462 g/L (Curves 1 and 2) and 304 g/L (3); H_2SO_4 concentration = 98 g/L (1) and 1 g/L (2).

Recovery of some metal impurities contained in technical niobium hydroxide has been studied under batch conditions. The results are presented in Table 1. It is seen from the data obtained that large amounts of niobium can be extracted into the PEG phase and separated from other elements in a PEG-$(NH_4)_2SO_4$-NH_4F-H_2SO_4-water system.

Niobium can be stripped from the PEG phase into an ammonium sulfate solution of different composition, or precipitated with aqueous ammonia or other precipitant directly from the extract. The alkalization of the extract results in the formation of the second salt phase containing $(NH_4)_2SO_4$ and some impurities coextracted with niobium. The niobium hydroxide filtrate can be calcined to obtain purified Nb_2O_5.

A niobium purification scheme is given in Figure 5. The scheme has been tested using multistep mixer-settler equipment.

Table 1. Extraction of metal impurities by 40% PEG from sulfate-fluoride solutions.

Metal impurity	Nb solution composition[1]	
	1	2
Zn	0.7	0.9
Co	1.1	1.0
Fe	0.5	0.6
Cr	1.8	1.6
Mn	—	2.5
Eu	—	0.6

[1]Solution composition (g/L): 1, C_{Nb} = 139, C_{NH4F} = 170, $C_{(NH4)2SO4}$ = 488; 2, C_{Nb} = 152, C_{NH4F} = 187, $C_{(NH4)2SO4}$ = 462, C_{H2SO4} = 98.

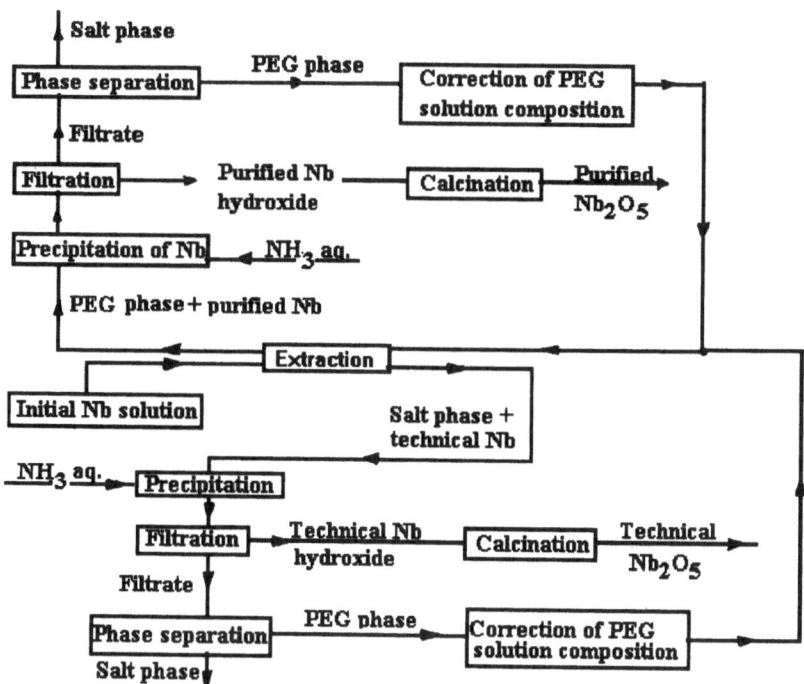

Figure 5. Scheme of niobium purification.

REFERENCES

1. P.-Å. Albertsson. "Partition of Cell Particles and Macromolecules," 3rd ed., Wiley, New York (1985).
2. H. Walter, D.E. Brooks, and D. Fisher (eds.), "Partitioning in Aqueous Two-Phase Systems. Theory, Methods, Uses, and Application to Biotechnology," Academic Press, Orlando, Florida (1985).
3. T.I. Zvarova (Nifant'eva), V.M. Shkinev, B.Ya. Spivakov, and Yu.A. Zolotov, Liquid extraction in the systems: aqueous solution of salt-aqueous solution of polyethylene glycol, *Doklady AN SSSR* 273:107 (1983).
4. T.I. Zvarova (Nifant'eva), V.M. Shkinev, G.A. Vorob'eva, B.Ya. Spivakov, and Yu.A. Zolotov, Liquid-liquid extraction in the absence of usual organic solvents: application of two-phase aqueous systems based on a water-soluble polymer, *Microchim. Acta* 3:449 (1984).
5. V.M. Shkinev, N.P. Molochnikova, T.I. Zvarova, B.Ya. Spivakov, B.F. Myasoedov, and Yu.A. Zolotov, Extraction of complexes of lanthanides and actinides with arsenazo III in ammonium sulphate-poly(ethylene glycol)-water two-phase system, *J. Radioanalyt. Chem.* 88:115 (1985).
6. S.N. Timasheff and G.D. Fasman (eds.), "Structure and Stability of Biological Macromolecules," Marcel Dekker, New York (1969).
7. R.D. Rogers, A.H. Bond, and C.B. Bauer, Aqueous biphase systems for liquid/liquid extraction of f-elements utilizing polyethylene glycols, *Sep. Sci. Technol.* 28:139 (1993).
8. T.I. Nifant'eva, V.M. Shkinev, B.Ya. Spivakov, Theoretical description of the distribution of metal ions in aqueous polymer extraction systems used in biotechnology, *in:* "Proceedings of 'Separations for Biotechnology,'" Reading, UK (1994).
9. T.I. Nifant'eva, V.M. Shkinev, B.Ya. Spivakov, and Yu.A. Zolotov, Metal extraction in two-phase aqueous systems polymer-polymer-salt-water, *Doklady AN SSSR* 308:879 (1989).
10. N.P. Molochnikova, V.M. Shkinev, and B.F. Myasoedov, Two-phase aqueous systems based on poly(ethylene glycol) for extraction separation of actinides in various media, *Solv. Extr. Ion Exch.* 10:697 (1992).
11. R.D. Rogers, A.H. Bond, and C.B. Bauer, Metal ion separations in polyethylene glycol-based aqueous biphasic systems, *Sep. Sci. Technol.* 28:1091 (1993).
12. T.I. Nifant'eva, V.M. Shkinev, B.Ya. Spivakov, and Yu.A. Zolotov, Liquid-liquid extraction of thiocyanate and halide complexes of metals in two-phase aqueous systems of poly(ethylene glycol)-salt-water, *Zh. Analyt. Chimii* 44:1368 (1989).

THE BEHAVIOR OF ACTINIDES IN TWO-PHASE AQUEOUS SYSTEMS BASED ON POLYETHYLENE GLYCOL

B.F. Myasoedov, N.P. Molochnikova, V.M. Shkinev, and
B.Ya. Spivakov

Vernadsky Institute of Geochemistry and Analytical Chemistry
Russian Academy of Sciences
Moscow 117975, Russia

INTRODUCTION

As one of the most effective analytical and technological separation techniques, liquid/liquid extraction is based on distribution of compounds between an aqueous solution and an organic solvent immiscible towards the former. The effectiveness of the process markedly depends on the energy difference between solvation and hydration of extracted compounds in the two phases. Water-soluble organic and inorganic agents of high hydration energy usually form poorly extracted complexes in traditional extraction systems. Meanwhile, a number of water-soluble organic reagents exhibit significant complexing ability and selectivity with respect to metal ions including lanthanides and actinides. Such reagents could substantially extend the sphere of applications of liquid-liquid extraction to separation and concentration of elements. Among reagents of this type are various derivatives of aminoacetic acids (so called complexones), a number of organic photometric reagents such as Arsenazo III and other complexones. Most water-soluble reagents do not usually form extractable complexes in conventional extraction systems due to the high hydration energy of the species to be extracted. These reagents can be used in heterogeneous aqueous systems with considerable amounts of water in both phases where the effect of hydration on transfer of the extracted complex from one phase to another is not significant. Such extraction systems use water-soluble polymers whose salting out from an aqueous electrolyte solution results in formation of a second liquid phase.[1] One of these is poly(ethylene glycol) (PEG), a cheap, easily available, and non-toxic polymer producing two-phase aqueous systems with various salts.[1,2] These systems are of interest to us from a practical point of view, since they contain no organic solvents which are usually volatile, explosive, and toxic.

Water is the single liquid component of the systems considered. That is why the term "solvent extraction" may not be used in this case and one should only use the term "liquid-liquid extraction" or "liquid-liquid distribution" as recommended by IUPAC.

Aqueous Biphasic Separations: Biomolecules to Metal Ions
Edited by R.D. Rogers and M.A. Eiteman, Plenum Press, New York, 1995

RESULTS AND DISCUSSION

Separations of actinides were conducted with systems using PEG (Loba Chemia) with molecular mass of 2000. Within these systems we studied separation of actinides in carbonate, sulfate, phosphate, rhodanide, and nitrate solutions in the presence of both organic and inorganic water-soluble complexing agents. Conditions for quantitative group extraction of actinides and for separation of these elements from U, Th, and lanthanides have been found.[3]

Extraction from Carbonate Solutions[4-6]

Separations of actinides in carbonate solutions are of great interest due to their extensive use in practical work.[7] We have studied in detail the distribution of actinides and lanthanides in K_2CO_3-PEG-water-complexone systems. Extraction of actinides into the PEG phase in the absence of complexing agents is not significant and it does not exceed 1% (Table 1). Addition of complexones results in a significant increase of distribution coefficients for actinides and lanthanides. Quantitative extraction occurs in the presence of Xylenol Orange (XO) and Alizarine Complexone (AC), while lower distribution coefficients are achieved for the complexes with Methylthymol Blue (MTB) and hydroxyphenylimino-N,N'-diacetic acid (HPIDAA). The best extracting agents for the elements studied (XO, AC, and MTB) belong to complexones of the phthalexone type containing several benzene rings. Aliphatic complexing agents do not increase, as a rule, the extraction of the elements and in some cases even suppress the extraction into the PEG enriched phase (Table 1). Differences in extracting power of complexones can be used for more effective separation of trivalent actinides and lanthanides.

Table 1. Distribution coefficients of Am and Eu in the system K_2CO_3-PEG (30% weight)-H_2O-complexone (0.02 M).[4]

Reagent	D_{Am}	D_{Eu}
——	0.006	0.002
Xylenol orange, XO	32.7	16.7
Alizarine complexone, AC	57.8	32.7
Hydroxyphenylimino-N,N'-diacetic acid, HPIDAA	4.8	6.0
Methylthymol blue, MTB	2.5	1.0
Hydroxyethyldimethylphosphonic acid, HEDMPA	0.003	0.01
Nitrilotriphosphonic acid, NTPA	0.02	0.06
Hydroquinonecomplexone, HMIDA	0.15	0.04
Hydroquinoneiminodimethylphosphonic acid, HMIDPA	0.004	0.006
1,3-Dihydroxyphenylmethyliminodiacetic acid, DHPMIAA	0.1	0.07
Thymolphthalexone, TP	0.05	0.09
Ethylenediaminotetraacetic acid, disodium salt, EDTA	0.04	0.06
Diethylenetriaminopentaacetic acid, DTPA	0.3	0.21

A more detailed study has concentrated on the extraction of americium as a function of various factors such as concentration of complexones and carbonate ions in the system and the nature of the alkali metal cation. As the concentration of a complexing agent increases, the extraction of americium into the PEG phase increases (Figure 1). A complexing agent concentration higher than ca. 10^{-2} M results in no increase in the extraction of the element. Potassium carbonate concentration slightly affects the extraction of actinide complexes. Effective extraction (~ 97%) of trivalent actinides and lanthanides is possible even from 40% K_2CO_3, i.e., from highly concentrated solutions (Figure 2). Americium is extracted from the sodium carbonate solution at markedly lower concentration of complexone (Figure 1). This confirms the effect of the nature of alkali metal on the behavior of actinides in the carbonate solutions and is due to the difference in salting-out effects of corresponding cations. Isolation of actinides from the sodium carbonate solutions is more complete and separation of actinides is more effective in the potassium carbonate solutions. Data on separation of actinides in the system: PEG-water-potassium carbonate in the presence of various complexones are listed in Table 2. The order of the extraction of transplutonium elements in this system is Es > Cf > Bk > Cm > Am. In contrast to the trivalent actinides and lanthanides, the extraction of uranium, neptunium, and plutonium by PEG solution in the presence of XO is not significant thus favoring their effective separation. A low-level extraction of actinides in the oxidation state +5 in the XO system has made it possible to find conditions for separation of americium-243 from its daughter decay product neptunium-239 in carbonate solutions. The stability of macroamounts of americium(V) with respect to the polymer and complexone, and the low rate of its extraction in the presence of AC have made it possible to separate macroamounts of americium(V) from curium in carbonate solutions. The α-spectra in Figure 3 show that there is practically no curium in the salt phase after extraction. It passes quantitatively into the PEG phase containing AC. The coefficient of americium and curium separation per one cycle of extraction is greater than 100.

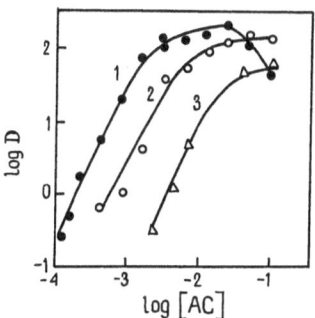

Figure 1. Americium(III) extraction as a function of AC (1,2) and XO (3) concentration in the systems: aqueous PEG solutions-potassium (2, 3) and sodium (1) carbonate solutions.

In two-phase aqueous systems based on PEG, a group separation of actinides and lanthanides is possible in the presence of two complexones, one (AC or XO) being the extractive agent and the other a masking agent. Masking agents were complexones of an aliphatic series containing both carboxyl and phosphoryl groups. The best results were demonstrated for the mixture of XO and hydroxyethyldiphosphonic acid (HEDPA).

In two-phase aqueous systems based on PEG, a quantitative extraction of trivalent and tetravalent actinides from highly concentrated carbonate solutions is possible in the presence of various complexones. It is also possible to separate these actinides from pentavalent and hexavalent elements with high separation coefficients. These systems are equal to or better than the best traditional extraction systems suitable for carbonate solutions.

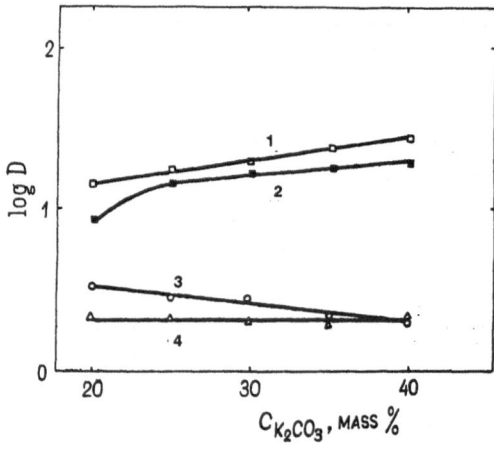

Figure 2. Americium(III) (1, 3, 4) and europium(III) (2) extraction as a function of K_2CO_3 concentration in the system: PEG-K_2CO_3-complexone (0.02 M): XO (1, 2), MTB (3), and HPIDAA (4).

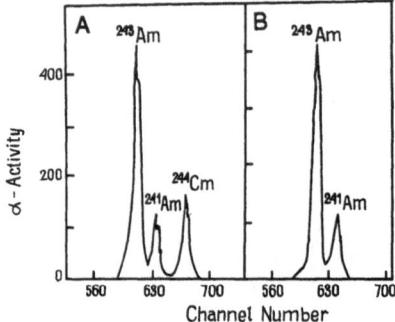

Figure 3. Alpha-spectra of americium and curium: initial solutions (A) and salt phases after extraction by PEG in the presence of AC (B).

Table 2. Separation of elements in the system PEG-K_2CO_3 (30% weight)-H_2O-complexone (0.02 M).

Complexone	AC			XO			MTB	
Element	D	Separated elements	ß	D	Separated elements	ß	D	Separated elements	ß
$^{253-254}$Es	88.7	TPE/Np	487	143.7	TPE/Np	679	0.61	TPE/Np	61.9
^{249}Cf	36.8	TPE/U	1364	69.6	TPE/U	3573	1.10	TPE/U	43.3
^{249}Bk	42.5			48.2	TPE/Pu	1306	1.42	TPE/Pu	37.2
^{243}Cm	67.6			47.4			0.95	TPE/Ce	18.8
^{241}Am	105.8			32.2			2.50		
^{239}Pu	13.4	Pu/Np	96	0.05			0.04		
^{237,239}Np	0.14	Pu/U	268	0.10			0.021		
^{233}U	0.05	Eu,Ce/Np	221	0.019	Eu,Ce/Np	168	0.030		
$^{152-154}$Eu	31.8	Eu,Ce/U	607	16.9	Eu,Ce/U	884	1.01	Eu/Ce	14.4
^{144}Ce	30.2			16.9	Eu,Ce/Pu	323	0.069		

Extraction from Sulfate Solutions[4,8-11]

PEG forms two-phase systems with ammonium sulfate solution in a wide range of salt concentrations. The system 15% PEG-14.4% weight of ammonium sulfate was chosen for the study of metal extraction. Figure 4 shows the extraction of the complexes of actinides and lanthanides with (curve 1) and without (curve 2) Arsenazo III as a function of equilibrium salt-phase pH. The distribution of these elements in the absence of Arsenazo III was less than 10% into the PEG-enriched phase. Quantitative extraction of actinides takes place only in the presence of Arsenazo III which acts as a complexing agent for the metal ions. The distribution coefficients of the f-elements increase with increasing concentration of Arsenazo III . As one can see from the data, extraction of actinides markedly depends on pH. A relatively acidic medium (pH > 2) is suitable for quantitative extraction of thorium(IV), plutonium(IV), and uranium(VI) in the order: Th > Pu > U. In the pH range of 4.0-4.5 there is quantitative extraction of the trivalent transplutonium elements and rare earths. The effectiveness of extraction, Bk > Cf > Cm > Am, is the same as the extraction of chelate compounds of these elements in traditional extraction systems. This system allows separation of thorium(IV) with a separation coefficient over 100 at pH 1.5, and plutonium(IV) with a separation coefficient of ca. 60 at pH 2.2 from trivalent TPE and REE. More than 90% of the thorium and plutonium are extracted into the PEG phase.

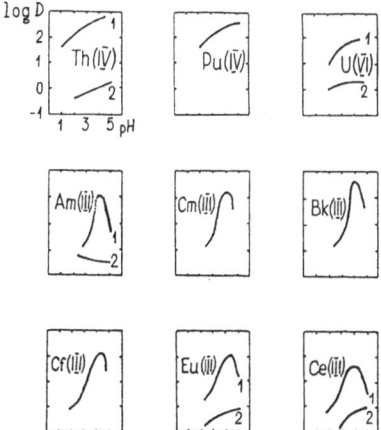

Figure 4. Distribution coefficients for metals in the system: PEG 2000 (15 mass %)-(NH$_4$)$_2$SO$_4$ (14.4 mass %)-H$_2$O as a function of equilibrium salt phase pH. Curves: 1) Extraction of metal ions with Arsenazo III (10^{-3} M); 2) Distribution of metal ions in the absence of Arsenazo III.

In the PEG-ammonium sulfate-water system, we have also studied the behavior of actinides in the presence of inorganic complexing agents such as potassium phosphotungstate, K$_{10}$P$_2$W$_{17}$O$_{16}$•nH$_2$O (PW). It has been found that anions of "unsaturated" heteropolytungstates pass into the PEG phase. A quantitative extraction of the trivalent and tetravalent actinides by the PEG solution occurs in the presence of PW concentrations of 5x10^{-4} M (Table 3). In the absence of potassium phosphotungstate, the extraction of actinides from ammonium sulfate solution into the PEG phase is insignificant with the exception of tetravalent elements.

Actinides in the oxidation states +5 and +6 are only slightly extracted by a PEG solution in the presence of PW, thus enabling one to separate the elements in different oxidation states. Separation coefficients are presented in Table 3. A high degree of purification of trivalent actinides from neptunium(V) has made it possible to develop a method for separation of americium-243 from the daughter neptunium-239 and to work out an isotope generator of neptunium-239. Figure 5 shows γ-spectra of an initial americium solution and a solution of ammonium sulfate and PEG after extraction with PW. From γ-spectra analysis it follows that it is only the americium that passes into the PEG phase while all of the neptunium remains in the salt phase. Neptunium-239 accumulated in the PEG phase may be washed out from the PEG solution by any salt solution that makes up a heterogeneous system with the PEG and contains no complexing agent.

Table 3. Separation of elements in the system PEG-$(NH_4)_2SO_4$-H_2O-PW.

Element	D Without PW	D [PW] = 5×10^{-4} M	Separated Elements	ß
Am(III)	0.16	137.4	Am, Cm/Np	1960
Cm(III)	0.16	120.2	Am, Cm/Pu	430
Eu(III)	0.25	12.6	Am, Cm/U	720
Ce(III)	0.25	63.7	Pu/Np	70
Pu(IV)	0.63	4.5	Pu/U	20
Zr(IV)	0.63	3.0	Ce/Np	910
Pa(V)	0.47	0.32	Ce/Pa	200
Np(V)	0.019	0.045	Ce/U	340
Am(V)	0.13	0.49	Eu/Np	180
U(VI)	0.18	0.30	Eu/U	60

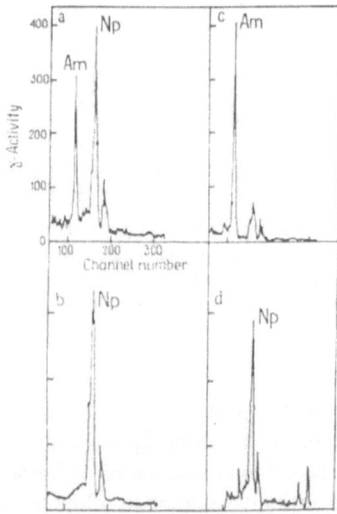

Figure 5. Separation of Am-243 and Np-239 in the system PEG-$(NH_4)_2SO_4$-H_2O-PW. γ-spectra: Americium initial solution (a), Salt phase after extraction (b), PEG phase after extraction (c), Ammonium sulfate solution with accumulated Np-239 back extracted from the PEG (d).

Extraction of actinides in the oxidation state +4 is slightly less than that of the trivalent elements and consequently, the separation coefficients for tetra-, penta-, and hexavalent actinides, shown in Table 3, are also lower. The data obtained suggest that two-phase aqueous systems based on PEG may be successfully used for isolation and separation of actinides, including americium in different oxidation states from sulfate solutions.

Extraction from Phosphate and Rhodanide Solutions[12]

PEG solutions form two-phase systems with mono- and disubstituted ammonium and sodium phosphates. We studied the extraction of americium by PEG from the solution of disubstituted ammonium phosphate in the presence of various complexing agents. The most complete extraction of americium is attained by PEG solutions containing either complexones of the phthalexone type such as XO and AC, or inorganic complexing agents such as potassium phosphotungstate.

The behavior of actinides in different oxidation states was most thoroughly studied in the system PEG-$(NH_4)_2HPO_4$-H_2O-PW (Table 4). Under these conditions tetra-, penta-, and hexavalent elements are extracted weakly.

Table 4. Separation of elements in the system PEG-$(NH_4)_2HPO_4$-H_2O-PW (0.001 M).

Element	D	Separated Elements	ß
Am(III)	6.7	Am/Np	444
Ce(III)	9.5	Am/Pu	27
Pu(IV)	0.25	Am/U	15
U(VI)	0.44	Ce/Np	527
Np(V)	0.018	Ce/Pu	38

The group separation of TPE and REE does not occur in the presence of rhodanide ions since the distribution coefficients for americium (D = 3.6) and europium (D = 2-3) in this system are close in value.

Extraction from Nitrate Solutions[13]

Two-phase aqueous systems based on PEG cannot be used for metal extraction from nitrate solution. Heterogeneous aqueous systems with nitrate solutions can be prepared only in the presence of two water-soluble polymers such as PEG-dextran sulfate or PEG-dextran (Table 5). The distribution coefficients for the metals are rather low in these systems in the absence of complexing agents. The distribution of the complexes of actinides with Arsenazo III, some phthalexones, and "unsaturated" heteropolytungstates has been studied. The distribution coefficients in the two-polymer systems are somewhat lower than in the PEG-salt systems (Table 6). Element recovery and separation can be improved by use of these systems in extraction and thin-layer chromatography. We used silica gel modified by dextran sulfate as a support and the PEG solution as a mobile phase. It has been shown that the two-phase aqueous systems based on two water-soluble polymers can be used for extraction and separation of actinides and for obtaining isotope generators of radioactive elements.

Table 5. Biphasic aqueous systems with two water-soluble polymers: PEG-dextran (DX) or PEG-dextran sulfate (DS).

Composition	Concentration	Part
DX or DS	20 % w/w	4
PEG	60 % w/w	2
KNO_3	2 M	1
H_2O (PW, Arsenazo III, complexones)		1

Table 6. Distribution coefficients of americium in the system PEG-DS.

Complexing agent	D_{Am}
-	0.08
PW	1.5-2.0
Arsenazo III	2.0-2.5
XO	0.3-0.5

CONCLUSIONS

Two-phase aqueous systems based on water-soluble polymers can be extensively used for concentration, separation, and extraction of actinides from various salt solutions. These systems hold much promise for solving a variety of practical problems related to using well-known water-soluble extraction agents.

REFERENCES

1. P.-Å. Albertsson. "Partition of Cell Particles and Macromolecules," Almquist and Wiksell, Stockholm (1971).
2. H. Walter, D.E. Brooks, and D. Fisher, eds., "Partitioning in Aqueous Two-Phase Systems. Theory, Methods, Uses, and Applications to Biotechnology", Academic Press, Orlando, Florida (1985).
3. N.P. Molochnikova, V.M. Shkinev, and B.F. Myasoedov, Two-phase aqueous systems based on polyethylene glycol for extraction separation of actinides in various media, *Solvent Extr. Ion Exch.* 10:687 (1992).
4. B.F. Myasoedov, N.P. Molochnikova, V.M. Shkinev, T.I. Zvarova, B.Ya. Spivakov, and Yu.A. Zolotov, Liquid-liquid extraction of complexes of actinides with water-soluble organic reagents in two phase salt-polyethylene glycol-water systems, *in*: "Proceedings of the International Symposium on Actinide and Lanthanide Separations," G.R. Choppin, J.D. Navratil, and W.W. Schulz, eds., World Scientific, Singapore (1985).
5. N.P. Molochnikova, V.M. Shkinev, B.Ya. Spivakov, Yu.A. Zolotov, and B.F. Myasoedov, Extraction of complexes of actinides and lanthanides in two-phase aqueous system: potassium carbonate-polyethylene glycol-water, *Radiokhimiya* 30:60 (1988).
6. N.P. Molochnikova, V.Ya. Frenkel, B.F. Myasoedov, V.M. Shkinev, B.Ya. Spivakov, and Yu.A. Zolotov, Extraction of actinides into aqueous PEG solutions from carbonate media in the presence of Alizarine Complexone, *Radiokhimiya* 29:330 (1987).
7. Z.K. Karalova, B.F. Myasoedov, T.I. Bukina, and E.A. Lavrinovich, Extraction and separation of actinides and lanthanides from alkaline and carbonate solutions, *Solvent Extr. Ion Exch.* 6:1109 (1988).

8. V.M. Shkinev, N.P. Molochnikova, T.I. Zvarova, B.Ya. Spivakov, B.F. Myasoedov, and Yu.A. Zolotov, Extraction of complexes of lanthanides and actinides with Arsenazo III in an ammonium sulfate-polyethylene glycol-water two-phase system, *J. Radioanal. Nucl. Chem., Articles* 88:115 (1985).

9. N.P. Molochnikova, V.Ya. Frenkel, B.F. Myasoedov, V.M. Shkinev, B.Ya. Spivakov, and Yu.A. Zolotov, Extraction of americium in different oxidation states in two-phase aqueous systems based on polyethylene glycol, *Radiokhimiya* 29:39 (1987).

10. N.P. Molochnikova, V.Ya. Frenkel, and B.F. Myasoedov, Extraction of actinides in two-phase water-polyethylene glycol-salt systems in the presence of potassium phosphotungstate, *J. Radioanal. Nucl. Chem., Articles* 121:409 (1988).

11. N.P. Molochnikova, V.Ya. Frenkel, and B.F. Myasoedov, Extraction of actinide complexes with phosphotungstate ions in two-phase aqueous system based on polyethylene glycol, *Radiokhimiya* 31:65 (1989).

12. N.P. Molochnikova, V.M. Shkinev, and V.Ya. Frenkel, Separation of actinides by using two-phase aqueous systems based on polyethylene glycol, *in:* "Proceedings of the International Conference Actinides '89," Nauka, Moscow (1989).

13. V.M. Shkinev, N.P. Molochnikova, and B.F. Myasoedov, Biphasic extraction systems based on two water-soluble polymers for actinide elements isolation, *in:* "Proceedings of All-Union Solvent Extraction Conference," Nauka, Moscow (1991).

AFFINITY PARTITIONING OF METAL IONS IN AQUEOUS BIPHASIC SYSTEMS: EXPERIMENTAL AND THEORETICAL ASPECTS

Roberto Guzmán and Carlos M. Téllez

Department of Chemical and Environmental Engineering
University of Arizona
Tucson, AZ 85721

INTRODUCTION

Even though the first report about aqueous biphasic systems appeared early in this century,[1] their regular application to the purification of biomolecules did not start until 1956 when the partial isolation of organelles from cell extracts using a poly(ethylene glycol) (PEG)/potassium phosphate system, was published by Albertsson.[2] Over the years, aqueous biphasic systems have been used successfully in several separation schemes and the number of applications has expanded tremendously. Most of the research has been in the separation and purification of proteins, cells, cell organelles, viruses, membrane fragments, and other biological materials.[3-7] In addition, aqueous biphasic systems have been used to characterize surface properties of biomolecules such as charge and hydrophobicity.[8]

In many cases, a major problem in partition-based separation methods is the minor differences in partition behavior for many biomolecules. Affinity-based interactions have been used in some instances in an attempt to increase the selectivity in the purification. This is usually accomplished by introducing a ligand into the system with affinity for the substance of interest, which itself partitions favorably to one of the phases. Thus, once binding occurs, the desired biomolecule preferentially distributes into the polymer-ligand rich phase. Normally, in protein purifications, the ligand is attached covalently to one of the phase forming polymers, which ensures its primary distribution to one phase in the system. Ligands attached to PEG have been extensively described, examples are PEG-linked dyes,[9] PEG-linked antibodies,[10] and PEG-linked long chained fatty acids.[11] Similar derivatives for dextran as well as some procedures to synthesize new affinity ligands have been elucidated by Harris[12] and Harris and Yalpani.[13] Chelating-PEG derivatives such as PEG-imino diacetic acid (PEG-IDA) with affinity for metal ions have been used as a tool to separate erythrocytes, heme containing proteins and phosphoproteins.[7,14,15] Correspondent to the experimental development of phase partitioning, mathematical theories have been described, essentially in all aspects of two-phase phenomena, including models from phase formation to charged solutes and protein partitioning with and without affinity interactions.

Aqueous Biphasic Separations: Biomolecules to Metal Ions
Edited by R.D. Rogers and M.A. Eiteman, Plenum Press, New York, 1995

101

Only until recently, aqueous biphasic systems have been used to extract metal ions from aqueous solutions. The first paper using these systems appeared in 1983,[16] and only a few more reports have been published since then.[17-21] Most of the relevant work so far has been with PEG/salt systems. In analogy to protein affinity partitioning, compounds that complex metal ions in solution have been used to enhance their distribution in a preferred polymer phase, usually the PEG rich phase. For the most part, organic agents such as dyes, crown ethers and inorganic ions such as thiocyanate, iodide and bromide have been the complexing agents of choice as the selective extractants.

In our laboratory, we have been synthesizing and using water-soluble metal affinity polymers (chelating-polymers) to explore their potential as tools to enhance the partitioning of metal ions in aqueous biphasic systems. Some of the experimental results with chelating-PEG derivatives in PEG/salt systems are presented. We also describe our efforts to model the behavior of this affinity metal ion partitioning approach. Our model is based on the thermodynamic criteria that govern the equilibrium among metal ions and ligand molecules in solution and their partitioning between phases.

BIPHASIC EXTRACTION OF METAL IONS

Up to now, all the work published in the literature in the partition of metal ions involve biphasic systems of PEG and an inorganic salt. In general, in the absence of any extractant or complexing agents, metal ions tend to partition favorably to the salt phase in an aqueous PEG/salt biphasic system. Zvarova et al.[16] reported the partition coefficients of several metal ions at two different pH values, and invariably, all of them had a value lower than one, ranging from 0.126 for sodium to 0.71 for indium. This result is somehow expected since the bottom phase (salt rich phase) would provide a better environment for the highly solvated and charged metal ions than the relatively hydrophobic polymeric top phase. Also measured were partition coefficients of some anions (Br$^-$, I$^-$, PO$_4^{3-}$, and SCN$^-$), and with the exception of the phosphate ion, all of them distribute favorably in the top phase. In fact this behavior, along with their ability to form complexes with some metal ions prompted their use as extractants in biphasic systems. A partition coefficient as high as 610 for technetium as pertechnetate anion in PEG/sulfate and PEG/carbonate systems without the addition of extractant agents of any kind, however, has been reported.[17] The explanation given for this behavior refers to the tightening of the water structure surrounding this large, univalent, unhydrated anion that could produce a salting out effect from the bottom phase. This suggests, as pointed out by Rogers et al.[17] that other metal ions with similar characteristics could show a similar behavior. Most studies of metal ion partitioning in aqueous two phase systems involve the addition of an organic extractant or an inorganic complexing agent that binds the metal ion and extracts it to the top phase.

Selective Extractants

Partitioning coefficients of 60 for Fe(III) and 1000 for Cu(II) have been reported using salts of the inorganic ion thiocyanate, values up to two orders of magnitude higher than the values obtained using iodide and bromide. Iodide complexes of some metals favorably partition in the top phase, ordinarily providing better results than their bromide counterparts. Rogers et al.[17] provide an extensive list of references where these and other inorganic and organic agents have been employed to extract metal ions using two phase systems.

Among the organic extractants reported in the literature, dyes constitute the most common types of complexing agents. In order for a particular dye to be successful in extracting metal ions, it needs to partition favorably to the top phase and needs to have a relatively high association constant with the metal ions. Zvarova et al.[16] suggests that the presence of aromatic rings in the molecule aids metal extraction. Most of the work so far has focused in the extraction of actinides and lanthanides. In most cases, a direct correlation has been observed between the amount of reagents and the extracted metals, although in some cases an upper limit is reached after which there is no improvement in the separation. From the performed experiments is not clear whether the limit is a feature of the system (i.e., composition, pH, dissociation constants, etc.) or due to the gradual depletion from the salt phase of free metal ion by the complexing agent.[16-19]

Crown ethers have been used to extract mainly Group 1 and 2 metal ions.[20] Among the crown ethers tested, 12-crown-4, 15-crown-5, and 18-crown-6, only the latter seems to induce partition coefficients higher than one. Rogers et al.[20] have observed that the nature of the salt in PEG/salt systems could be the difference between a successful extraction or no separation at all. The system PEG/NaOH has rendered the best results over PEG systems with ammonium sulfate and potassium carbonate. The authors have associated this behavior with the total amount of phase forming salt concentration in the system. Since NaOH requires the least amount of salt to form the two phases with PEG, this system produces the highest partition coefficients, in the presence and absence of extractant.

In a study involving the cations Na, Ca, Rb, Sr, Cs, and Ba in a PEG/NaOH system with 15-crown-5 and 18-crown-6 as extractants, a relationship between the presence of NO_3^- and the partition coefficients of the ions was found.[21] Extraction was markedly low with either crown ether when no nitrate salt was present in the system. However, when sodium nitrate was gradually added, a linear increase in the partition coefficient occurred. Improved values were higher by up to one order of magnitude except for sodium. In an attempt to gain a better understanding about the processes involved in metal ion partitioning, Rogers et al.,[20] have studied what the influence of the hydration enthalpy of a cation is on its distribution between the two phases. They suggest an inverse correlation between the enthalpy of hydration and partition coefficient for a given cation.

Although the partition coefficients reported for crown ethers, in the extraction of metal ions from aqueous solutions, are not as high as with other techniques, they have proved to be a viable alternative for some cations that are not extracted with the usual reagents. It is evident that more research needs to be performed to understand the fundamental factors involved in the partitioning of cations in these systems.

Modified Polymer-Chelators

Several modification procedures for PEG have been well documented in the last decade.[5,22,23] Most of them were directed towards the synthesis of ligands to be used in affinity partitioning of proteins. With the introduction of immobilized metal ion affinity chromatography (IMAC), several chelators for metal ions have been synthesized for column work on agarose matrices.[24] Resembling this separation scheme, IDA and other similar compounds, have been attached to PEG molecules and used as a tool in immobilized metal affinity partitioning of proteins.[14,25]

The extracting properties of these ligands are based on the chelating effect that some of their individual atoms display by acting as electron donors. Oxygen and nitrogen atoms in these molecules form coordination bonds with the metal ion in solution producing a metal chelate. Chelates are much more stable than a metal complex due to the loss in free energy when a ring is formed. The IDA ligand is tridentate and can form a double five-membered ring chelate with hexacoordinate metal ions, while TED (tris-carboxymethylated ethylene diamine) is pentadentate and capable of forming four five-membered rings.[26] These ligands work better with transition metal ions, which have several possible coordination sites and act as electron acceptors in the presence of these type of extractants.

Several of these same PEG-chelators have been synthesized in our laboratory and used for affinity biphasic separations of metal ions such as Cu(II), Co(II), and Ni(II) from aqueous solutions.[27,28] The metal affinity ligands shown in Figure 1 include (a) IDA-PEG, (b) L-aspartic acid polyethylene glycol (L-Asp-PEG), (c) TED-PEG, and (d) carboxymethylated (tris(2-aminoethyl) amine) polyethylene glycol (Cm-TREN-PEG).

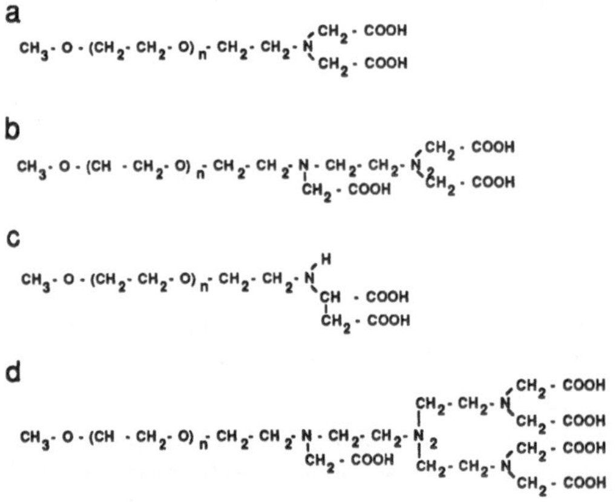

Figure 1. Structures of PEG-chelating derivatives used in affinity partitioning of metal ions.

In these systems the efficiency of extraction depends on the stability of the complex between the metal ion and the modified PEG, and the relative distribution of the ligand between the phases. Ligand partition coefficient values, in general were higher than 50. PEG-IDA and PEG-TED were found to be the best affinity chelators among all ligands tested, with the former performing slightly better than the latter. In general, metal partition coefficients increased more than two orders of magnitude compared to values obtained when no ligand was added to the system.

In an initial attempt to describe this affinity partitioning of metal ions, we derived a mathematical model that can predict complex formation constants as well as partition coefficients of metals and ligands in both phases.

MODELING AQUEOUS TWO PHASE PARTITIONING

In order to fully benefit from the multiple advantages of aqueous two phase partitioning, several investigators have proposed models to predict phase formation and the behavior of a number of molecules in these systems. Phase separation in polymer-polymer phase systems has been described in terms of the attractive interactions between molecules of the same polymer and the mutual repulsion between molecules of unlike polymers. Those interactions dominate the free energy of mixing of the system, and overcome the entropy gain by free mixing of the molecules in solution, since polymers have large molecular weights, and therefore large surface area.[29] Prediction of phase separation has in general been described by thermodynamic models based on the Gibbs free energy of mixing and statistical mechanical approaches.

A measure of the suitability of the separation of a particular molecule in a given two phase system is its partition coefficient (K) described as the ratio of the concentration in the top to the bottom phases. The success of aqueous two phase systems lies in the fact that most biomolecules would partition favorably into one phase, or its partition coefficient can be modified by changing some conditions in the system. Empirical evidence has shown that K

is a function of several variables including the choice of phase polymers, their molecular weight, phase composition, system ionic composition, system pH, and temperature among others.[30] Additionally, some physical characteristics of the molecule, such as charge, hydrophobicity, size, and concentration, play an important role in the partition process.[31]

Modeling Protein Partition

The Flory-Huggins theory[32,33] and the Edmond-Ogston model[34] have been used as the basis of some protein partitioning models. King et al.[35] proposed a model based on a modified theory of Edmond-Ogston to describe the partitioning of proteins in dilute solutions in polymer/polymer systems. Under these conditions, when the concentration of the protein approaches zero, and is much smaller than the concentration of both phase forming polymers, the protein partition coefficient (K_p) is given by:

$$\ln K_p = a_{2p}(m_2'' - m_2') + a_{3p}(m_3'' - m_3') + \frac{z_p F(\Phi'' - \Phi')}{RT} \tag{1}$$

where m_2 and m_3 are the corresponding concentrations of the polymers and a_{ij} is the interaction coefficient of species i with species j. The last term considers the effect of an electrical potential Φ between the phases, caused by the uneven partition of some salts added to buffer the system. The model does not explicitly consider the interactions between ions and polymers but provides an attempt to take into account electrostatic effects in these systems. The model qualitatively predicts the partitioning of lysozyme, bovine serum albumin, and α-chymotrypsin in a PEG/dextran two phase system.

Brooks et al.[36] developed a lattice model by extending the Flory-Huggins theory applicable to protein partitioning. They consider a four-component system containing water, and three polymeric solutes, one of which is the protein. Under the same assumption of relative low concentration of the protein, the partition coefficient (K) is given by:

$$\ln K = P_4 \left\{ [\phi_{11} - \phi_{12}](1 - \chi_{1/2}) + [\phi_{21} - \phi_{22}]\left(\frac{1}{P_2} - \chi_{2/4}\right) + [\phi_{31} - \phi_{32}]\left(\frac{1}{P_3} - \chi_{3/4}\right) \right\} \tag{2}$$

where ϕ_{ij} is the volume fraction of component i in phase j. $\chi_{i/j}$ is the interaction energy (Flory-Huggins interaction parameter) for a segment of species i interacting with a segment of species j. P_i is the number of lattice sites occupied by a single molecule of species i. Subindex 4 refers to the protein, 1 to the water, and 2 and 3 to the phase forming polymers.

The model qualitatively describes some of the empirical trends of protein partitioning, such as: favorable protein partitioning into one of the phases proportional to protein molecular weight and to the polymer concentration difference between the phases.

Similarly, Diamond and Hsu[8,37] developed a relationship for the protein partition coefficient (K):

$$\frac{\ln K}{w_1'' - w_1'} = A^* + b^*(w_1'' - w_1') \tag{3}$$

where A^* is a function of protein and phase forming polymer molecular weight, protein charge, protein-water, protein-polymer, and protein-protein interaction parameters, pH, electrostatic potential difference between the phases, and salt type and concentration. The term b^* is a function of the protein molecular weight and charge, the polymer-water interaction parameters, and electrostatic potential differences between the phases and salt type and concentration as well. The quantity $(w''_I-w'_I)$ is known as the PEG concentration difference between the phases. This model correlates well with protein partition data from the literature in polymer/polymer and polymer/salt two phase systems. It clearly presents the effect of the phase forming polymer on the partition of proteins. Other models have been developed in similar form and are presented in more detail in a different section of this book.[30,38-40]

In a different approach, Albertsson[3,41] proposed that the partition coefficient of a biomolecule in an aqueous two phase system could be represented as a function of additive contributions to give:

$$\ln K = \ln K_o + \ln K_{elec} + \ln K_{hfob} + \ln K_{biosp} + \ln K_{conf} \tag{4}$$

where *elec*, *hfob*, *size*, *biosp*, and *conf* refer to electrochemical, hydrophobic, size, biospecific and conformational contributions to the partition coefficient from both the protein structural properties and the surrounding environment of the system, respectively. Another form to express that partition coefficient is then:

$$\ln K = \ln K_{environment} + \ln K_{structure} \tag{5}$$

The environment conditions influencing the partitioning include: salt type and concentration, pH, type, molecular weight, and concentration of the phase forming polymers, possible polymer derivatives present with charged, affinity, or hydrophobic groups, temperature, and gravity. The structural properties of the protein include molecular weight, primary, secondary, tertiary, and quaternary structure, net charge, hydrophobicity, and other surface properties.[8]

Taking some of these factors into account, Eiteman and Gainer[42] proposed a model for the partitioning of molecules based on their relative hydrophobicity. The following relation was derived for the partition coefficient (K):

$$RT \ln K = (\alpha_p + \Delta f) \Delta \omega_2 \tag{6}$$

where α_p is a phase constant that depends on the phase system, but is independent of the concentration of the phase components. $\Delta \omega_2$ is the concentration difference between the phases of the main component in the top phase. Δf is the total relative hydrophobicity of a large molecule composed of several constituent molecules bound by condensation reactions. The model is based on the linearity of the partition coefficient of a homologous series of compounds with the free energy required to transfer the monomeric unit of the series between the phases.[43] The relative hydrophobicity is normalized by selecting a solute which has been arbitrarily assigned a value of $\Delta f = 0$, and data from this solute are then used to calculate the phase constant.

The model successfully predicts partition coefficients of several aminoacids, peptides, esters, and alcohols at system conditions that render them neutral.[44,45] However, when the solutes are charged, i.e., at high pH, the predictions deviate from the experimental data.

In order to deal with the problem of the charge in some solutes partitioning in aqueous two phase systems, Eiteman[46] developed a correlation that allows the calculation of partition coefficients of multicharged solutes when the partition coefficient of the neutral species is known, and is given by:

$$K = k \, \frac{x_o' + \sum_{i=1}^{m} x_{i+}' + \sum_{j=1}^{n} x_{j-}'}{x_o'' + \sum_{i=1}^{m} x_{i+}'' + \sum_{j=1}^{n} x_{j-}''} \qquad (7)$$

where m and n are the maximum number of positive and negative charges that the solute can have depending on the pH of the solution respectively. x_{i+} and x_{j-} are the mole fractions of the solute bearing i positive charges and j negative charges respectively. Single and double superscripts represent top and bottom phases in that order. x_o is the mole fraction of the neutral solute and k is the partition coefficient of the uncharged molecule.

The mole fractions are expressed in terms of the activities and activity coefficients, and these in turn are obtained from equilibrium constants (K_b and K_c):

$$K_{bi} = \frac{a_{(i-1)} a_{H^+}}{a_{i^+}} \qquad (8)$$

$$K_{cj} = \frac{a_{j^-} a_{H^+}}{a_{(j-1)^-}} \qquad (9)$$

where K_{bi} and K_{cj} are for positively and negatively charged solutes.

In order to calculate the partition coefficient of the multicharged solute, only the knowledge of the equilibrium constants, pH, and partition coefficient of the neutral molecule are needed. The model qualitatively describes the partition behavior of several polypeptides in a polymer/salt two phase system.

Modeling Affinity Protein Partitioning

The first affinity partition model that appeared in the literature was proposed by Flanagan and Barondes.[47] They consider the change in Gibbs free energy (ΔG) for the transfer of 1 mole of the complex ligand-biomolecule from the top to the bottom phase as the sum of the individual free energy changes in the process. The total Gibbs free energy is given by the energy change for dissociating the complex in the top phase, transferring the dissociated species across the interface (ligand and protein separately), and then reassociating the complex in the bottom phase.

$$\Delta G = \Delta G_1 + \Delta G_2 + \Delta G_3 + \Delta G_4 \qquad (10)$$

where ΔG_i is the change in free energy for each process described above. In terms of an equilibrium constant (Q), this becomes:

$$\Delta G_i = -RT \ln Q_i \qquad (11)$$

For the phase transfer operation, Q_i represents a partition coefficient of the involved species (K_i), and for the association and reassociation steps, Q_i is the equilibrium binding constant (k_a). After expressing the equilibrium constants in terms of the species concentrations, the partition coefficient of the complex ligand-biomolecule (K_P) can be described by the following equation, for the special case of a biomolecule interacting with two ligands:

$$K_P = K_{Po} \left(\frac{1 + l_{totT} / K_{a,T}}{1 + l_{tot,T} / K_{aB} K_L} \right)^2 \tag{12}$$

where K_{Po} is the partition coefficient of the biomolecule with no ligand present, and l_{tot} is the total ligand concentration in any phase. K_L is the partition coefficient of the free ligand and K_a is the dissociation constant of the enzyme biomolecule-ligand complex. Subscripts T and B refer to the top and bottom phases respectively.

In 1985, Brooks et al.[36] proposed a more general model, that did not assume equal dissociation constants for the complex in both phases. They expressed the partition coefficient of the biomolecule (K_P) in the usual manner, as:

$$K_P = \frac{M_{totT}}{M_{tot,B}} \tag{13}$$

where M_{tot} is the total concentration of biomolecules for that phase. One of the assumptions is that the biomolecule has n possible binding sites and that at any given conditions it can have i number of them occupied by the ligand. They express this total concentration of biomolecule (M_i) as:

$$\sum_i M_i = M_o (1 + K_d L)^n \tag{14}$$

where K_a is the dissociation constant, L is the total concentration of ligand, M_i is the concentration of biomolecule with i sites occupied by ligand, and M_o is the total biomolecule concentration in the system. Substituting the above equation in the partition coefficient expression for each phase, and rearranging, the following expression is obtained:

$$K_m = K_o \frac{(1 + K_a^T L^T)^n}{(1 + K_a^B L^B)^n} \tag{15}$$

where K_o is the partition coefficient of the biomolecule with no ligand present. If the ligand concentration in both phases is very high, then all of the biomolecules will be saturated and the above equation reduces to the Flanagan and Barondes model.

Cordes et al.[48] have developed an improved model by making a balance of total affinity ligand and biomolecule, adding up free and all bound species. They consider as well, a biomolecule that has n binding sites and their derivation leads to the equation:

$$K_P = K_{Po} \left(\frac{1 + l_{tot,T} / K_{a,T}}{1 + l_{tot,T} / K_{aB} K_L} \right)^n \tag{16}$$

This expression, however, did not fit some of their own experimental data. They proposed instead, the inclusion of a second dissociation constant when the biomolecule had two binding sites. The partition coefficient is then given by:

$$K_P = K_{Po} \frac{1 + 2l_{totT} / K_{a1,T} + L_{tot,T}^2 / K_{a1,T}K_{a2T}}{(1 + L_{tot,T} / K_{a1,T}K_L)^2} \tag{17}$$

Cordes *et al.*[48] have extended their derivation for cases when the biomolecule has four and six different binding sites. Clark and Sandler[4] also proposed to treat affinity partition adding contributions to the partition coefficient. The expression for the partition coefficient of a biomolecule, using Brooks' model to represent the affinity contribution, becomes:

$$\ln K_m = \ln K_o + \ln \left(\frac{(1 + K_a'L')^n}{(1 + K_a''L'')^n} \right) + \frac{zF\Delta\Phi}{RT} \tag{18}$$

where K_o is the partition coefficient in the absence of affinity ligand and charge in the biomolecule. The last term accounts for charge effects. z is the net charge of the molecule, F is the Faraday constant and $\Delta\Phi$ is the electrical potential difference between the phases. Similarly, other models have been developed.[49-51]

Modeling Immobilized Metal Affinity Protein Partitioning

In 1990, Suh *et al.*[52,53] presented a model for metal affinity partitioning of proteins. It is an extension of the theory by Cordes *et al.*[48] to account for inhibition of the binding sites in the protein to which the affinity ligand binds. They consider hydrogen ions as inhibitors of the free base form of the imidazole ring in histidine residues responsible for the protein binding to the immobilized metal ion in metal affinity partitioning. The theory considers the protein having n independent binding sites, and the same association constant between the metal ion and the surface-exposed histidines for each site. In their model, the pK_a of the imidazole ring is the same for all histidine residues as well.

By assuming that the pH is the same in both phases, and that the free ligand concentration is greater than the concentration of the ligand bound to protein, the resulting equation for the partition coefficient (K) based on a mass balance for the protein, becomes:

$$\ln \left(\frac{K}{K_o} \right) = n\ln \left[\frac{1 + K_a' \dfrac{R + 1}{\left(R + \dfrac{1}{K_M} \right)} M_{tot} + K_H H}{1 + K_a'' \dfrac{R + 1}{(K_M R + 1)} M_{tot} + K_H H} \right] \tag{19}$$

where K_o is the partition coefficient of the protein in the absence of affinity ligand, R is the volume ratio of the top and bottom phases, M_{tot} is the total metal concentration, H is the hydrogen ion concentration, K_M is the partition coefficient of the ligand, and K_a and K_H are the intrinsic association constants for the metal and hydrogen ion, respectively. Again, prime and double prime denote the top and bottom phases, respectively.

The expression was used to model experimental data for the partition coefficients of horse myoglobin, whale myoglobin, and cytochrome c from *Candida krusei*, at different hydrogen and affinity ligand concentrations. They used the complex polyethylene glycol-iminodiacetic acid-Cu (Cu-IDA-PEG) in a PEG/Dextran aqueous two phase system.

The model has been extended to consider other aminoacids besides histidine that have affinity for the immobilized metal ion, such as cysteine or tryptophan.[54] In this model the protein can have m binding sites with a metal association constant K_b, and hydrogen ion association constant $K_{H,b}$, for cysteine for example; and n sites with a metal association constant K_a and hydrogen ion association constant $K_{H,a}$, for histidine. In that case the resulting expression is:

$$\ln\left(\frac{K}{K_o}\right) = n\ln\left[\frac{1 + K_a'M' + K_{H,a}H}{1 + K_a''M'' + K_{H,a}H}\right] + m\ln\left[\frac{1 + K_b'M + K_{H,b}H}{1 + K_b'M' + K_{H,b}H}\right] \tag{20}$$

EXPERIMENTAL - AFFINITY PARTITIONING OF METAL IONS

Partition experiments using PEG-chelate derivatives were performed in PEG/salt biphasic systems. The aqueous two phase system used in these experiments were prepared from stock solutions of PEG 8000, PEG/IDA-PEG, PEG/TED-PEG, and the salt (Na_2SO_4). The PEG 8000 stock solution was prepared by dissolving 120 grams of PEG in 180 grams of deionized water to give a final concentration of 40% (w/w). The IDA-PEG stock solution was prepared dissolving 6 grams of IDA-PEG with 114 grams of PEG 8000 in 180 grams of deionized water to give a final concentration of 40% (w/w) in total polymer and 2% (w/w) in IDA-PEG. The stock solution of TED-PEG was prepared in identical manner to give the same final concentration. For the preparation of salt stock solution, 80 grams of anhydrous sodium sulfate, reagent grade, were dissolved in 420 grams of deionized water to give a final concentration of 16% (w/w) in salt.

Solutions of metal ions were prepared using $CuSO_4 \cdot 5H_2O$, $CoSO_4 \cdot 7H_2O$, and $NiSO_4 \cdot 6H_2O$ from Fisher Chemicals, all metal salts were reagent grade. Solutions with a final concentration of 3000 ppm in metal ion were prepared for the experiments. Dilutions of the metal solutions were made to give concentrations of 2000 ppm, 1400 ppm, 1000 ppm, and 400 ppm of metal ion.

Aqueous two phase systems were prepared mixing 3.5 grams of polymer stock solutions with 5.0 grams of sodium sulfate stock solution. The phases were formed in plastic conical test tubes. One gram of metal solution (with different concentration of metal for each case) was added to the freshly prepared two phase system and 0.5 gram of deionized water was incorporated to give a total weight for the system of 10 grams. After preparation of the systems the plastic test tubes were inverted 50 times and left for 20 minutes to reach equilibrium. The final concentration of the systems prepared in this manner were 8% (w/w) in salt, 14% (w/w) in total polymer, and metal concentration depending on the dilution added to the system. Metal dilutions of 3000 ppm, 2000 ppm, 1400 ppm, 1000 ppm, and 400 ppm will produce solutions with metal concentrations of 0.326 g/mL, 0.217 g/mL, 0.152 g/mL, 0.109 g/mL, and 0.043 g/mL, respectively. After phase separation, the combined volume of top and bottom phases was 9.2 mL. The top phase volume was 4.3 mL and the bottom phase was 4.9 mL. The pH of top and bottom phases was measured after addition of metal solution for each system prepared, and was 6.0 for both top and bottom phases. The experimental partition coefficients were measured at room temperature once the systems had reached equilibrium as described by Aguiñaga et al.[27]

The partition coefficients for these systems with several metal ions is presented in Figures 2 and 3, where the distribution of metal ions in the top phase is presented as a function of the total concentration in the system, in the presence of chelating polymer. The difference in the logarithm of partition coefficients ($\Delta \ln K$) is defined as the partition coefficient of the metal ion with no affinity ligand in the system, subtracted from the partition coefficient of the metal ion in the presence of the affinity ligand, ($\ln K - \ln K_o$).

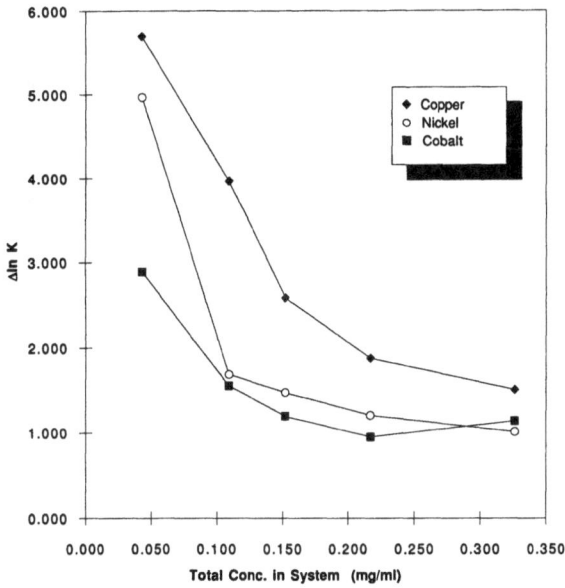

Figure 2. Partition coefficient behavior of different metal ions in aqueous PEG/Na₂SO₄ systems using the modified affinity chelating-polymer PEG-IDA.

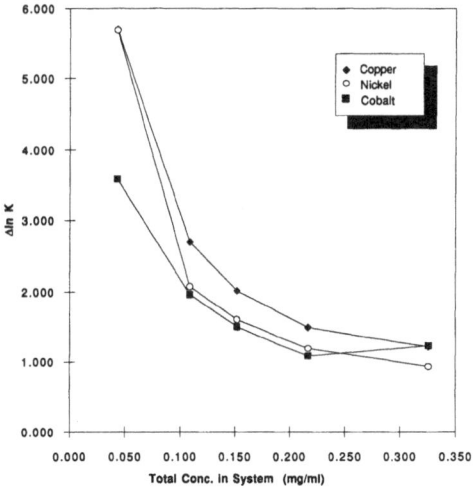

Figure 3. Partition coefficient behavior of different metal ions in aqueous PEG/Na₂SO₄ systems using the modified affinity chelating-polymer PEG-TED.

Based on the performed experiments, copper is highly extracted by PEG-IDA compared to nickel and cobalt. Figures 2 and 3 clearly show that the difference of the logarithm of partition coefficients for each chelator-PEG is higher at lower concentrations of metal ion in the system. This may be due to the saturation of the chelating sites available in the polymer. At low concentrations of metal ions in the system, enough chelating sites are available in the polymer to bind all of them, however, as the metal ion concentration increases, less metal binding sites are available causing a decrease in the metal partition coefficient.

MATHEMATICAL MODELING OF AFFINITY PARTIONING OF METAL IONS

The present model was developed in an effort to describe the partitioning of metal ions in aqueous biphasic systems using affinity polymer-chelators. The model has been used to calculate complex formation constants between metal ions in solution and the modified chelating agents, as well as partition coefficients of both of these species. It is based on the thermodynamic criteria that govern the equilibrium between the phases and was developed using IDA-PEG as the model affinity ligand.[55]

The model considers that only totally dissociated IDA-PEG can bind to the free metal ion, and that the molar ratio metal/ligand in the complex is 1. The equilibrium reactions for the molecule IDA-PEG (LH_2) are given by:

$$LH_2 \leftrightarrow H^{\cdot} + LH^{-}$$

$$LH^{-} \leftrightarrow H^{\cdot} + L^{-2} \tag{21}$$

and

$$L^{-2} + M^{\cdot} \leftrightarrow ML \tag{22}$$

The last equation is governed by the metal-ligand complex dissociation constant (K_{fML}). The partition coefficient of the complex is represented as K_{ML}.

Since the affinity polymer-chelator is a covalently bound complex of a molecule of PEG and a molecule of an affinity ligand, the PEG fraction of the complex is usually much larger than that of the affinity ligand. It can be considered, to a first approximation, that the partition coefficient of the various forms of the PEG-ligand complex are dominated by the PEG fraction. It can be further assumed that all of these partition coefficients are approximately equal. Under these circumstances, the following relationships between the apparent dissociation constants ($K_{h,1}$ and $K_{h,2}$) of the ligand in the top and bottom phases can be derived.

$$K_{h,1}^{T} \approx K_{h,1}^{B} \cdot 10^{(pH^{B} - pH^{T})} \tag{23}$$

$$K_{h,2}^{T} \approx K_{h,2}^{B} \cdot 10^{(pH^{B} - pH^{T})} \tag{24}$$

where pH^{T} and pH^{B} are the pH in the top and bottom phases respectively.

In the experiments performed with this particular two phase system, the pH was found to be the same for both phases. It has been assumed, therefore, that the aforementioned dissociation constants are the same in both phases. Other investigators, have proposed alternative expressions when a difference in the pH of the phases exists.[44]

The model is based on the thermodynamic equations that govern the equilibrium among metal ions, hydrogen ions, and ligand molecules in solution, and their partitioning between the phases. Along with the equilibrium equations for the protonation/deprotonation of the ligand and for the formation of the ligand-metal complex, mass balances of the ligand and metal ions provide the needed equations that describe the phenomena involved. The resulting equations are solved simultaneously, and the parameters (K_{fML}, K_{ML}, $K_{h,1}$, and $K_{h,2}$) are fitted to the experimental data using the Hook-Jeeves method of optimization.

Figures 4 and 5 show the outcome of increasing the metal ion concentration on the partition coefficient, and a simulation of the effect of the pH in the partition of the metal ion, respectively.

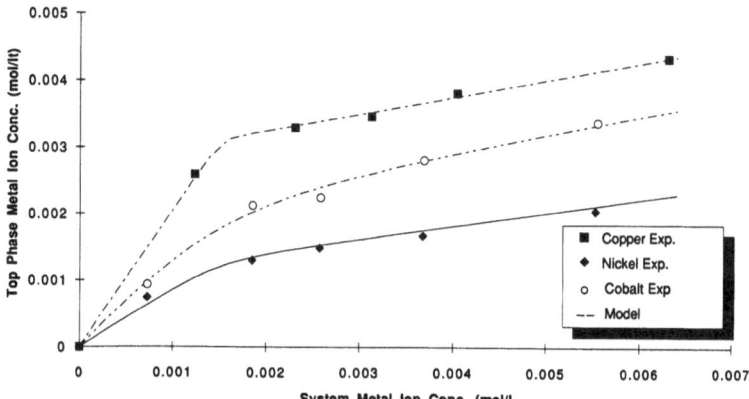

Figure 4. Distribution of metal ion concentration in the top phase as a function of total concentration of a given metal ion in the system in the presence of the chelating-polymer IDA-PEG.

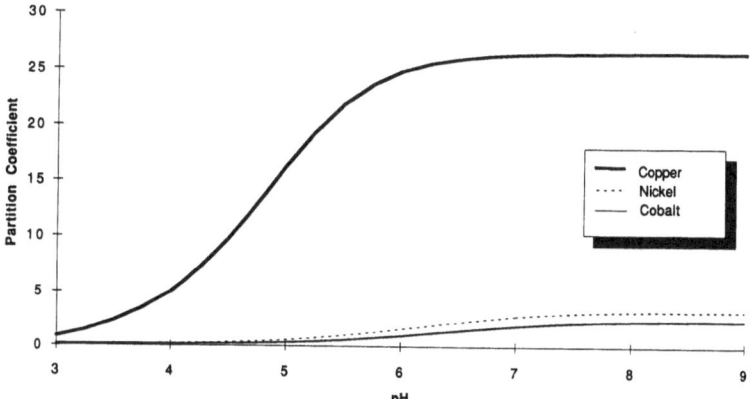

Figure 5. Predicted partition coefficients of metal ions as a function of pH in the presence of the chelating-polymer PEG-IDA.

The model explains the mechanism by which the partition coefficient of the metal ion increases when an affinity ligand is used in an aqueous two phase system. In general, the metal ion is displaced by hydrogen ions when the pH is low. At a pH when all the ligand molecules are totally protonated, the partition coefficient of a metal ion will be practically independent of the concentration of affinity ligand, and will have a value very close to the partition coefficient when no ligand is present in the two phase system. As the pH increases, more ligand molecules are able to bind free metal ion, increasing its partition coefficient. However, the maximum value of the metal ion partition coefficient cannot be greater than the partition coefficient of the ligand itself, and this value is achieved when all metal ions have been chelated by the ligand.

The findings from this modeling approach can be applied directly to the recovery of metal ions from aqueous solutions that use affinity or pseudo affinity ligands, like modified polymers, dyes, etc. These findings as well, might be of interest in the understanding of immobilized metal affinity partitioning of biomolecules. Previous reports[52,53] considered only the interactions between the protein and the complexed metal ion, but they have not considered those between the metal ion and the affinity ligand, specifically, neglecting the effect of pH on the stability of the complex.

CONCLUSIONS

The area of metal ion aqueous partitioning with biphasic systems is experiencing a systematic degree of interest, even though until now only results using PEG and inorganic salts with and without added components have been reported. In order to expand in this area of metal ion partitioning, extensive research is needed, including characterization and development of new systems, new extractants, and studies of basic solution chemistry and mathematical modeling. Currently there is an extensive number of variables that apply to metal ion partitioning in biphasic systems. At this point it seems as if metal ion partitioning is following in the foot steps of protein partitioning, not only in terms of experimentation, but also in terms of theoretical developments. In analogy to the use of specific ligands attached to PEG molecules in protein affinity partitioning, and more specific in metal affinity partitioning, chelating modified polymers ligands have been used to enhance the partitioning of metal ions such as Cu(II), Ni(II), and Co(II) in aqueous biphasic systems. The use of affinity ligand-polymers is quite promising and at this point we can only hypothesize, based on the theory of partitioning, on possible improvements that can be made to increase the efficiency of this extraction process. For instance, the higher the molecular weight of the modified PEG molecule and the lower the molecular weight of the phase forming polymer, the higher the partition coefficient of the ligand would be. New modified polymers can be synthesized with different functional groups at each end adding flexibility to their use. Additional research is needed to determine binding constants between modified chelators and specific metals, as well as knowledge on the influence of several other variables, such as pH, ionic strength, binding constants, and specificity of the ligands towards given metal ions, and the evaluation of competitive extraction when several transition metal ions are present in solution.

Among the metal ion extraction techniques discussed, each has advantages and disadvantages, however, they have been employed under different conditions. Dyes have been used with actinides and lanthanides, crown ethers have been used additionally with Group I and II cations, and modified PEG ligands with transition metal ions. Aside from the relative efficiency of each process, perhaps the most significant difference of importance in the scaling up process is that dyes may be expensive and potentially toxic. In addition, they are not specific and would extract most of the metal ions present in solution, while crown ethers and modified PEG could be tailored to be specific if needed.

The mathematical model for affinity partitioning of metal ions presented here, although preliminary, has been used to satisfactorily describe the partition behavior of the metal ions Cu(II), Ni(II), and Co(II), and association constants between affinity ligands and metals. The model shows the relevant effect that the pH and dissociation constants have on the partitioning of metal ions in aqueous biphasic systems.

REFERENCES

1. P.-Å. Albertsson, History of aqueous polymer two phase systems, *in*: "Partitioning in Aqueous Two Phase Systems: Theory, Methods, Uses and Applications to Biotechnology," pp. 1-10, H. Walter,D.E. Brooks, and D. Fisher, eds., Academic Press, New York (1985).
2. P.-Å. Albertsson, Chromatography and partition of cells and cells fragments, *Nature* 177:771 (1956).
3. P.-Å. Albertsson, "Partition of Cell Particles and Macromolecules," 3rd. ed., Wiley-Interscience, New York, (1986).
4. W.M. Clark and S.I. Sandler, Affinity partitioning and its potential in biotechnology, *Sep. Sci. Technol.* 23:761 (1988).
5. "Partitioning in Aqueous Two Phase Systems: Theory, Methods, Uses and Applications to Biotechnology," H. Walter,D.E. Brooks, and D. Fisher, eds., Academic Press, New York (1985).
6. W. Müller, Partitioning of nucleic acids, *in*: "Partitioning in Aqueous Two Phase Systems: Theory, Methods, Uses and Applications to Biotechnology," pp. 227-266, H. Walter,D.E. Brooks, and D. Fisher, eds., Academic Press, New York (1985).
7. H. Goubran-Botros, A.O. Birkenmeier, G. Koppershlager, and M.A. Vijayalakshmi, Immobilized metal ion affinity partitioning of cells in aqueous two-phase systems: erythrocytes as a model, *Biochim. Biophys. Acta* 1074:69 (1991)
8. A.D. Diamond and J.T. Hsu, Aqueous two-phase systems for biomolecule separation, *in*: "Advances in Biochemical Engineering/Biotechnology," Vol 47., pp. 89-135, A. Fiechter, ed., Springer-Verlag, Berlin (1992).
9. G. Koppershlager and G. Johansson, Affinity partitioning with polymer-bound cibacron blue F3G-A for rapid large-scale purification of phosphofructokinase from baker's yeast, *Anal. Biochem.* 124:117 (1981).
10. K.A. Sharp, M. Yalpani, S.J. Howard, and D.E. Brooks, Synthesis and application of a poly(ethylene glycol)-antibody affinity ligand for cell separations in aqueous polymer two phase systems, *Anal. Biochem.* 154:110 (1986).
11. V.P. Shanbhag and G. Johansson, Interaction of human serum albumin with fatty acids, role of anionic groups studied by affinity partition, *Eur. J. Biochem.* 93:363 (1979).
12. J.M. Harris, Laboratory synthesis of polyethylene glycol derivatives, *J. Macromol. Sci.* C-25:325 (1985).
13. J.M. Harris and M. Yalpani, Polymer-ligands used in affinity partitioning and their synthesis, *in*: "Partitioning in Aqueous Two Phase Systems: Theory, Methods, Uses and Applications to Biotechnology," pp. 589-626, H. Walter,D.E. Brooks, and D. Fisher, eds., Academic Press, New York (1985).
14. G.E. Wuenschell, E. Naranjo, and F.H. Arnold, Aqueous two phase metal affinity extraction of heme proteins, *Bioprocess Eng.* 5:199 (1990).
15. B.H. Chung and F.H. Arnold, Metal-affinity partitioning of phosphoproteins in PEG/dextran two phase systems, *Biotechnol. Lett.* 13:615 (1991).
16. T.I Zvarova, V.M. Shkinev, G.A. Vorob'eva, B.Ya. Spivakov,and Yu.A.Zolotov, Liquid-liquid extraction in the absence of usual organic solvents: application of two phase aqueous systems based on a water soluble polymer, *Mikrochim. Acta* III:449 (1984).
17. R.D. Rogers, A.H. Bond, and C.B. Bauer, Metal ion separations in polyethylene glycol-based aqueous biphasic systems, *Sep. Sci. Technol.* 28:1091 (1993).
18. N.P. Molochnikova, V.M. Shkinev, and B.F. Myasoedov, Two phase aqueous systems based on poly(ethylene glycol) for extraction separation of actinides in various media, *Solvent Extr. Ion Exch.* 10:697 (1992).
19. R.D. Rogers, A.H. Bond, and C.B. Bauer, Aqueous biphase systems for liquid/liquid extraction of f-elements utilizing polyethylene glycols, *Sep. Sci. Technol.* 28:139 (1993).
20. R.D. Rogers, C.B. Bauer, and A.H. Bond, Novel polyethylene glycol-based aqueous biphasic systems for the extraction of strontium and cesium, *Sep. Sci. Technol.* in press (1994).
21. R.D. Rogers, A.H. Bond, and C.B. Bauer, The crown ether extraction of group 1 and 2 cations in polyethylene glycol-based aqueous biphasic systems at high alkalinity, *Pure Appl. Chem.* 65:567 (1993)
22. A.F. Buckman and M. Morr, Functionalyzation of poly(ethylene glycol) and monomethoxy poly(ethylene glycol), *Makromol. Chem.* 182:1379 (1981).
23. A.F. Buckman, M. Morr, and M.R. Kula, Preparation of technical grade polyethylene glycol (PEG) (Mr 20,000)-N6-(2-Aminoethyl)-NADH by a procedure adaptable to large scale synthesis, *Biotechnol. Appl. Biochem.* 9:269 (1987).
24. J. Porath, IMAC-immobilized metal ion affinity based chromatography, *Trends Anal. Chem.* 7:254 (1988).

25. G. Birkenmeier, M.A. Vijayalakshmi, T. Stigbrand, and G. Kopperschläger, Immobilized metal ion affinity partitioning (IMAP), a method for metal protein interaction and partitioning of proteins in aqueous two phase systems, *J. Chromatogr.* 539:267 (1991).

26. J. Porath, Immobilized metal affinity chromatography, *Protein Expression Purif.* 3:263 (1992).

27. P.A. Aguiñaga, "Polymer Modification with Metal Chelates and Their Application to the Recovery of Metals and Biocompounds from Aqueous Solutions," Masters thesis, University of Arizona (1992).

28. P.A. Aguiñaga, C.M. Téllez, and R. Guzmán, Synthesis of soluble modified chelating polymers with affinity for metal ions, *Synthetic Communications* submitted (1994)

29. N.L. Abbot, D. Blankschtein, and T.A. Hatton, On protein partitioning in two-phase systems, Bioseparation 1:195 (1990).

30. J.N. Baskir, T.A. Hatton, and U.W. Suter, Protein partitioning in two phase aqueous polymer systems, *Biotechnol. Bioeng.* 34:541 (1989).

31. P.-Å. Albertsson, Interaction between biomolecules studied by phase partition, *Methods Biochem. Anal.* 29:1 (1983).

32. P.J. Flory, Thermodynamics of high polymer solutions, *J. Chem. Phys.* 9:660 (1941).

33. M.L. Huggins, Solutions of long chain compounds, *J. Chem. Phys.* 9:440 (1941).

34. E. Edmond and A.G. Ogston, An approach to the study of phase separation in ternary aqueous systems, *Biochem. J.* 109:569 (1968).

35. R.S. King, H.W. Blanch, and J.M. Prausnitz, Molecular thermodynamics of aqueous two phase systems for bioseparations, *AIChE J.* 34:1585 (1988).

36. D.E. Brooks, K.A. Sharp, and D. Fisher, Theoretical aspects of partitioning, *in*: "Partitioning in Aqueous Two Phase Systems: Theory, Methods, Uses and Applications to Biotechnology," pp. 11-84, H. Walter,D.E. Brooks, and D. Fisher, eds., Academic Press, New York (1985).

37. A.D. Diamond and J.T. Hsu, Correlation of protein partitioning in aqueous polymer two phase systems, *J. Chromatogr.* 513:137 (1990).

38. D. Forciniti, C.K. Hall, and M.R. Kula, Temperature dependence of the partition coefficient of proteins in aqueous two phase systems, *Bioseparation* 1:227 (1991).

39. N.L. Abbot, D. Blankschtein, and T.A. Hatton, Protein partitioning in two phase aqueous polymer systems. 1. Novel physical pictures and a scaling thermodynamic formulation, *Macromol.* 24:4334 (1991).

40. J.N. Baskir, T.A. Hatton, and U.W. Suter, Thermodynamics of the separation of biomaterials in two phase aqueous polymer systems: effect of the phase forming polymer, *Macromol.* 20:1300 (1987).

41. P.-Å. Albertsson, General aspects of aqueous two phase partition, *in*: "Separations Using Aqueous Two Phase Systems," pp. 3-5, D. Fisher and I.A. Sutherland, eds., Plenum, New York (1989).

42. M.A. Eiteman and J.L. Gainer, Predicting partition coefficients in polyethylene glycol-potassium phosphate aqueous two phase systems, *J. Chromatogr.* 586:341 (1991).

43. B.Y. Zaslavsky, L.M. Miheeva, and S.V. Rogozhin, Parameterization of hydrophobic properties of aqueous polymeric biphasic systems and water-organic solvent systems, *J. Chromatogr.* 212:13 (1981).

44. M.A. Eiteman and J.L. Gainer, Prediction of partition coefficients for peptides in aqueous two phase systems, *in*: "Chromatographic and Membrane Processes in Biotechnology," pp. 323-333, C.A. Costa and J.S. Cabral, eds., Kluwer Academic, Dordrecht (1991).

45. M.A. Eiteman and J.L. Gainer, A correlation for predicting partition coefficients in aqueous two phase systems, *Sep. Sci. Technol.* 27:313 (1992).

46. M.A. Eiteman, Predicting partition coefficients of multicharged solutes in aqueous two phase systems, *J. Chromatogr.* 668:21 (1994)

47. S.D. Flanagan and J.H. Barondes, Affinity partitioning, *J. Biol. Chem.* 250:1484 (1975).

48. A. Cordes, J. Flossdorf, and M.R. Kula, Affinity partitioning: development of mathematical model describing behavior of biomolecules in aqueous two phase systems, *Biotechnol. Bioeng.* 30:514 (1987).

49. A. Carlson, Factors influencing the use of aqueous two phase partitioning for protein purification, *Sep. Sci. Technol.* 23:785 (1988).

50. J.N. Baskir, T.A. Hatton, and U.W. Suter, Thermodynamics of the separation of biomaterials in two phase aqueous polymer systems: comparison of lattice model to experimental data, *J. Phys. Chem.* 93:969 (1989).

51. J.P. Chen and J.T. Jen, Affinity partition of acid proteases in aqueous two-phase systems: modeling and protein purification, *J. Chem. Eng. Japan* 26:669 (1993).

52. S.-S. Suh and F.H. Arnold, A mathematical model for metal affinity protein partition, *Biotechnol. Bioeng.* 35:682 (1990).

53. S.-S. Suh, M.E. Van Dam, G.E. Wuenschel, S. Plunkett, and F.H. Arnold, Novel metal affinity protein separations, *in*: "Protein Purification: From Molecular Mechanisms To Large-Scale Processes," ACS Symposium Series, vol. 427, pp. 139-149, M.R. Ladish, R.C. Willson, C.C. Painton, and S.E. Builder, eds., American Chemical Society, Washington, DC (1990).

54. J.W. Wong, R.L. Albright, and N.L. Wang, Immobilized metal ion affinity chromatography (IMAC): chemistry and biochemistry applications, *Sep. Purif. Methods* 20:49 (1991).

55. C.M. Téllez, H. Cabezas, Jr., and R. Guzmán, Modeling affinity partitioning of metal ions in aqueous polymer/salt two-phase systems, *in preparation* (1994).

PURIFICATION OF BIOMOLECULES USING TEMPERATURE-INDUCED PHASE SEPARATION

Folke Tjerneld[1], Patricia A. Alred[1], Richard F. Modlin[2], Antoni Kozlowski[3], and J. Milton Harris[3]

[1]Dept. of Biochemistry
University of Lund
S-221 00 Lund, Sweden
[2]Dept. of Biological Sciences
[3]Dept. of Chemistry
University of Alabama in Huntsville
Huntsville, AL 35899

INTRODUCTION

There is a group of polymers which phase separate in water solution when the temperature is increased. These polymers have a lower critical solution temperature (LCST), which is also called the cloud point of the system. Above the critical temperature the polymers are not soluble in water. Examples of thermo-separating polymers are ethylene oxide (EO)/propylene oxide (PO) random copolymers and hydrophobically modified cellulose derivatives.[1,2] A water phase and a liquid, concentrated polymer phase are formed at temperatures above the cloud point of the EO/PO copolymer. Figure 1 shows the cloud point diagram for the EO/PO random copolymer Ucon 50-HB-5100. Factors determining the cloud point are EO/PO ratio, molecular weight and salt concentration. Non-ionic surfactants, such as Triton X-114, also have a LCST in water, and this property has been used for isolation of membrane proteins.[3]

Aqueous two-phase systems are widely used for separation and purification of biomolecules.[4,5] These systems are formed in water solutions of two incompatible polymers, such as poly(ethylene glycol) (PEG) and dextran. PEG is enriched in the upper phase and dextran in the lower phase. Both phases contain 80 to 95% water. Cells, cell particles and biomolecules, such as proteins, DNA and RNA, will distribute between the upper and lower phases as a result of several factors, including type and molecular weight of polymers, salt, pH, size and surface properties.[4-6]

Aqueous Biphasic Separations: Biomolecules to Metal Ions
Edited by R.D. Rogers and M.A. Eiteman, Plenum Press, New York, 1995

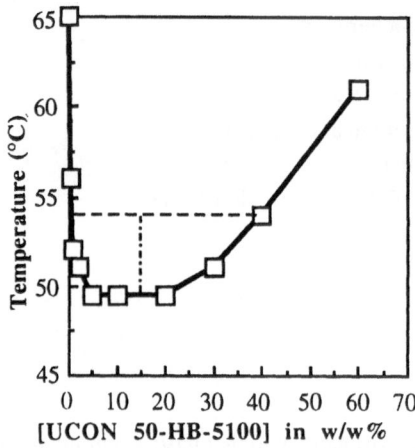

Figure 1. Cloud point diagram for the system UCON 50-HB-5100/water.[2] A tie-line has been drawn at 54°C to show the composition of the two phases at equilibrium at this temperature.

A new type of aqueous two-phase system, composed of an ethylene oxide/propylene oxide random copolymer as upper phase and either dextran or hydroxypropyl starch as lower phase polymer has been developed.[7-9] Partitioning in aqueous two-phase systems can in this way be combined with temperature-induced phase separation. A biomolecule can first be partitioned into a phase containing a thermo-separating polymer. This phase can then be removed and the temperature increased above the cloud point of the polymer. This will result in a new phase separation between a water phase and a polymer phase. The biomolecule can be separated from the polymer by partitioning to the water phase and the polymer can be reused. Both protein[7-9] and steroid[10,11] purification have been studied using this technique.

PROTEIN PURIFICATION

The random copolymer Ucon 50-HB-5100 (Union Carbide, New York, NY) has a composition of 50% ethylene oxide (EO), 50% propylene oxide (PO) and has a cloud point of 50°C. Its molecular weight is 4000.[2] A two-step protein purification method has been developed where Ucon/dextran or Ucon/hydroxypropyl starch aqueous polymer two-phase systems are used in a first step.[7] The composition of the system is selected so that the target protein is partitioned to the Ucon phase and the cell debris to the dextran or starch phase. The Ucon phase is removed and isolated in a separate container. The cloud point of Ucon is lowered to 40°C by addition of salt. The temperature is increased above the cloud point of Ucon which leads to the formation of a new two-phase system with an upper water phase and a lower Ucon phase. The high polymer concentration (>40%) in the lower phase leads to exclusion of the protein from this phase, and the protein is partitioned totally to the water phase. The polymer can be recycled back to the first extraction step. The target protein is obtained in the water phase free of polymer.

In order to lower the temperature of phase separation from water an ethylene oxide/propylene oxide random copolymer composed of 20% EO and 80% PO ($EO_{20}PO_{80}$) was synthesized.[8] This polymer has a cloud point of 18°C and makes it thus possible to perform temperature-induced phase separation close to room temperature. Ethylene oxide/propylene oxide random copolymers with different cloud points can now be obtained from Shearwater Polymers, Inc., Hunstville, AL.

Purification of Intracellular Enzymes

The method described above has been used for purification of intracellular enzymes from bakers' yeast.[7] Purification of 3-phosphoglycerate kinase (3-PGK) from yeast homogenate is shown in Table 1. The phase system used was Ucon 50-HB-5100 (Union Carbide, New York, NY) and dextran T500 (Pharmacia, Uppsala, Sweden) with 20% added homogenate. The extraction process began with a primary Ucon/dextran two-phase system at room temperature. In this initial step 3-PGK had a partition coefficient of 0.56, while K for total protein was 0.16. After separation, the upper Ucon-rich phase was removed and the temperature increased to 40°C. In the new two-phase system, the target enzyme partitioned completely into upper water phase, leaving the lower Ucon phase free of contamination. This Ucon phase was recovered. The polymer could be recycled and used in a second extraction of the original dextran lower phase. Total recovery of 3-PGK from both extractions in the combined upper water phases obtained by temperature-induced phase separation at 40°C was 56.8%, with a purification factor of 5.2.[7]

Table 1. Purification of 3-phosphoglycerate kinase from yeast homogenate using recycled Ucon to perform a second extraction. System is 6% Ucon 50-HB-5100, 3% dextran T500, 0.01M TEA-HCl buffer, pH 8.0, 0.2M sodium sulfate, and 20% yeast homogenate. K is the partition coefficient, C_t/C_b, and the distribution ratio $G = K \times (V_t/V_b)$.

Step		K (22°C)	G (22°C)	Purification Factor[1] (40°C)	% Units Recovered in Water Phase
1st extraction	3-PGK	0.56	1.57	6.6	36.6
	Protein	0.16	0.45	-	-
2nd extraction	3-PGK	0.41	0.90	3.8	20.2
	Protein	0.20	0.44	-	-
Total for both extractions[2]				5.2	56.8

[1]The specific activity of 3-phosphoglycerate kinase in the homogenate was 0.853 units/mg protein, equivalent to a purification factor of 1.

[2]Calculated for combined water phases.

Temperature-Induced Phase Separation at Ambient Temperature

The same process as above was performed using a two-phase system composed of EO20PO80, cloud point 18oC, and dextran T500 (Tables 2a and 2b). The primary two-phase partitioning was performed at 4°C and the temperature-induced phase separation at 24°C.[8] Partition of glucose-6-phosphate dehydrogenase (G6PDH), hexokinase and 3-phosphoglycerate kinase from yeast homogenate was determined. K values in the original EO20PO80 /dextran system at 4°C were 0.18 for G6PDH, 0.53 for hexokinase, 0.98 for 3-PGK and 0.05 for total protein content.

The EO20PO80 polymer is hydrophobic and partitioning in the EO20PO80/dextran system reflects hydrophobicity of proteins. 3-PGK is partitioned most to the EO/PO phase and is most hydrophobic of the proteins studied. The majority of water-soluble proteins are hydrophilic and therefore total protein strongly partitions to the bottom phase. In order to increase enzyme recovery, the volume of upper phase was increased. After temperature-induced phase separation the total yields in the upper water phase at 24°C were 50% for G6PDH, 67% for hexokinase and 72% for 3-PGK (Table 2b). The very low partition coefficient for total protein meant that fairly good purification factors were obtained: 5.3, 15.3 and 15.4 for the three enzymes, respectively.[8]

Table 2a. Partition of glucose-6-phosphate dehydrogenase, hexokinase and 3-phosphoglycerate kinase from yeast homogenate. Primary phase system: 8.5% $EO_{20}PO_{80}$, 2.0% dextran T500, 0.02 M sodium phosphate buffer, pH 7.0, and 20% yeast homogenate. K and G values at 24°C are not given as there was no detectable enzyme activity or protein in lower phase at this temperature. Volume ratio (V_t/V_b) at 4°C was 4.0.

Sample	K(4°C)	G(4°C)
Protein	0.05	0.20
G6PDH	0.18	0.72
Hexokinase	0.53	2.12
3-PGK	0.98	3.92

Table 2b. Purification of glucose-6-phosphate dehydrogenase, hexokinase and 3-phosphoglycerate kinase from yeast homogenate. Primary phase system: 8.5% $EO_{20}PO_{80}$, 2.0% dextran T500, 0.02M sodium phosphate buffer, pH 7.0, and 20% yeast homogenate. Y= % yield of enzymes. PF = purification factor at 24°C[1].

Sample	G6PDH		Hexokinase		3-PGK	
	PF	Y	PF	Y	PF	Y
Raw homogenate	1	-	1	-	1	-
After centrifugation	2.4	71	2.1	64	3.2	96
Upper phase - 4°C	4.2	52	13.0	69	15.7	80
Upper phase - 24°C	5.3	50	15.3	67	15.4	72

[1]The specific activity of an enzyme (units per mg protein) in the homogenate was equivalent to a purification factor of 1.

Purification Scheme Using Temperature-Induced Phase Separation

A scheme for enzyme purification using temperature-induced phase separation is shown in Figure 2. This purification process allows for recycling of copolymer solution, and recovery of enzyme in a water/buffer solution at 24°C. Original EO20PO80 /dextran system is formed at 4°C. In this system, it has been shown that most of total protein will partition to lower dextran phase, and upper copolymer phase can be removed and placed at 24°C.[8] The upper water phase formed after temperature increase contains the target enzyme, and is virtually free of copolymer. The low temperature (24°C) at which this copolymer phase separates from water eliminates any heat denaturation of sensitive enzymes and also does not require addition of salt in order to obtain phase separation at a low temperature.

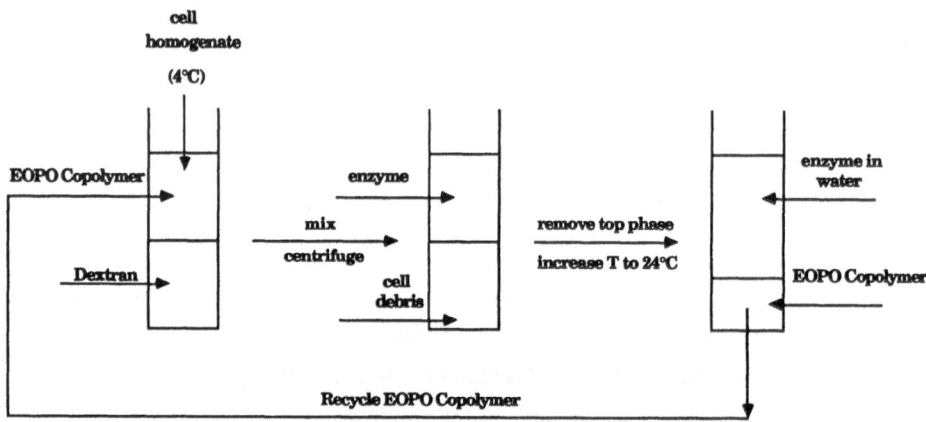

Figure 2. Enzyme purification scheme using aqueous two-phase partitioning at 4°C and temperature-induced phase separation at 24°C, with recycling of the $EO_{20}PO_{80}$ copolymer.

AFFINITY PARTITIONING

Temperature-induced phase separation has been utilized where the affinity ligand Procion Yellow HE-3G was covalently bound to Ucon for purification of enzymes.[9] In this technique an initial affinity partitioning step is performed in a Ucon/dextran aqueous two-phase system containing Ucon-ligand and cell extract. This step is similar to earlier work where PEG with bound affinity ligands has been used for affinity partitioning.[12] The Ucon phase is then isolated in a separate container. After temperature increase above the cloud point of Ucon the enzyme can be recovered in the water phase and the Ucon-ligand in the Ucon phase. Ucon-ligand plus Ucon can be recycled for a renewed extraction (see Fig. 3).

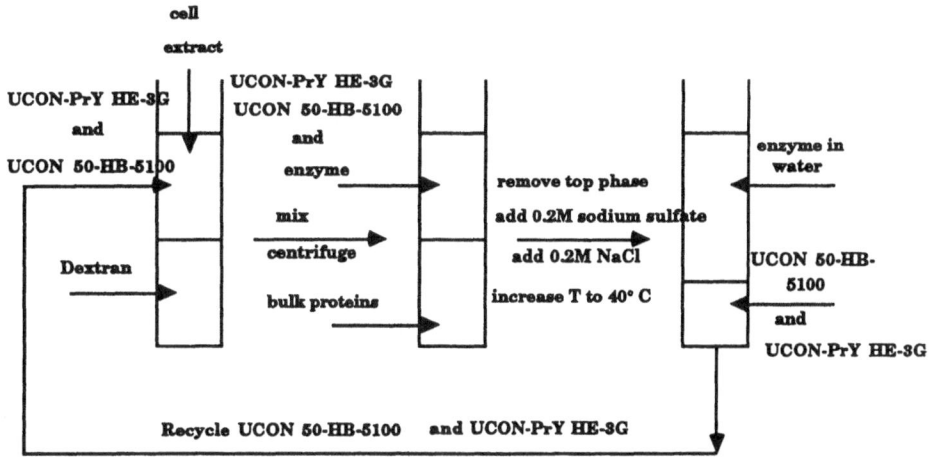

Figure 3. Enzyme purification scheme using affinity extraction of target enzyme with UCON-Procion Yellow HE-3G. Temperature-induced phase partitioning is used for enzyme recovery and recycling of UCON 50-HB-5100 and UCON-ligand.

Affinity Partitioning Studied with a Pure Enzyme

Table 3 shows the results of affinity partitioning and temperature-induced phase separation with pure glucose-6-phosphate dehydrogenase. G6PDH was partitioned in a Ucon/dextran phase system with 0.2% Ucon-Procion Yellow HE-3G.[9] In this phase system G6PDH was partitioned to the top Ucon-rich phase (K=4.5). After phase separation had occurred at 22°C, the Ucon-containing phase was removed. Sodium sulfate was added to this Ucon-phase to a concentration of 0.2M. This solution was placed in a water bath at 40°C for 15 minutes in order to induce phase separation. By the salt addition and temperature increase the enzyme was dissociated from the affinity ligand. In the new phase system G6PDH was totally partitioned to the upper water-salt phase. Enzyme was recovery was 88% in the water phase. The Ucon-Procion Yellow partitioned to the lower, Ucon-rich phase (K_L=0.06) and 77% could be recovered along with unmodified Ucon. There was no enzyme present in the lower Ucon and Ucon-Procion Yellow phase at 40°C.

Two partitionings are achieved by performing temperature-induced phase separation on the phase which contains the Ucon-ligand-enzyme complex. The enzyme is partitioned 100% to the water phase due to the steric exclusion from the concentrated polymer phase. The Ucon with bound ligand is partitioned to the Ucon phase because of the thermodynamic force which favours phase separation between Ucon and water at this temperature.

Table 3. Affinity partition of glucose-6-phosphate dehydrogenase with recovery of Ucon 50-HB-5100-Procion Yellow HE-3G. System is 5.1% Ucon 50-HB-5100, 7% dextran T500, 0.2% Ucon-Procion Yellow HE-3G, and 0.01M sodium phosphate buffer, pH 7.0. The amount of glucose-6-phosphate dehydrogenase was 34 units. K values at 40°C are for the partition between the water and Ucon phase formed by the increase in temperature.

$K_E(22°)$[1]	$K_E(40°)$	$K_L(40°)$[2]	% G6PDH recovered in water phase	% Ucon-PrY recovered in Ucon phase
4.51	>100	0.06	88.0%	77.1%

[1]K_E is the partition coefficient for G6PDH.

[2]K_L is the partition coefficient for Ucon-Procion Yellow HE-3G.

Affinity Purification of Enzyme from Yeast Extract

Glucose-6-phosphate dehydrogenase was purified from yeast extract in a Ucon/dextran aqueous two-phase system using 0.2% Ucon-Procion Yellow HE-3G (Table 4).[9] In the initial phase system the enzyme was extracted by the Ucon-ligand to the top phase (K=12). The bulk proteins were partitioned to the bottom phase (K=0.32). The upper phase was isolated in a separate container. Sodium sulfate and sodium chloride were added, both at 0.2M, and the temperature was raised to 40°C. In the new two-phase system formed at 40°C the enzyme was recovered in the water phase with a yield of 79% and a purification factor of 4.2. The partition coefficient for the enzyme in the water/Ucon phase system was >100. Ucon-Procion Yellow was recovered in the lower Ucon phase with a yield of 85%. No protein could be detected in this Ucon phase.

Table 4. Purification of glucose-6-phosphate dehydrogenase from yeast extract. System is 6.3% Ucon 50-HB-5100, 9% dextran T40, 0.2% Ucon-Procion Yellow HE-3G, 0.02M sodium phosphate buffer, pH 7.0, and 5.7% yeast extract. K values at 40°C are for the partition between the water and Ucon phase formed by the increase in temperature.

Sample	K (22°C)	K (40°C)	Purification Factor (40°C)	% Recovered at 40°C
G6PDH	12.4	>100	4.2	78.8[1]
Protein	0.32	>100	-	-
Ucon-PrY	24.6	0.32	-	84.6[2]

[1]Recovered in water phase at 40°C.

[2]Recovered in Ucon phase at 40°C.

PARTITION AND PURIFICATION OF ECDYSTEROIDS IN
UCON/DEXTRAN AND UCON/HYDROXYPROPYL STARCH SYSTEMS

Partitioning behavior of ecdysone (α-ecdysone) and 20-hydroxyecdysone (β-ecdysone) was determined in an aqueous two-phase system composed of Ucon 50-HB-5100 and dextran T500.[10] Ecdysteroids are a group of polyhydroxylated steroids derived from cholesterol. When compared to presently used techniques which use high concentrations of organic solvents to extract and purify the ecdysteroids, with normal yields of 50% to 80%, aqueous two-phase partitioning combined with temperature-induced phase separation offers a simple and inexpensive technique for ecdysteroid purification.

Ecdysone and 20-hydroxyecdysone are unique among steroids because they are both water soluble. These molecules are identical in structure except for substitution of a hydroxyl group for a hydrogen at position 20 on 20-hydroxyecdysone. This substitution results in different hydrophobicities for these two molecules, with ecdysone being the most hydrophobic (Figure 4).

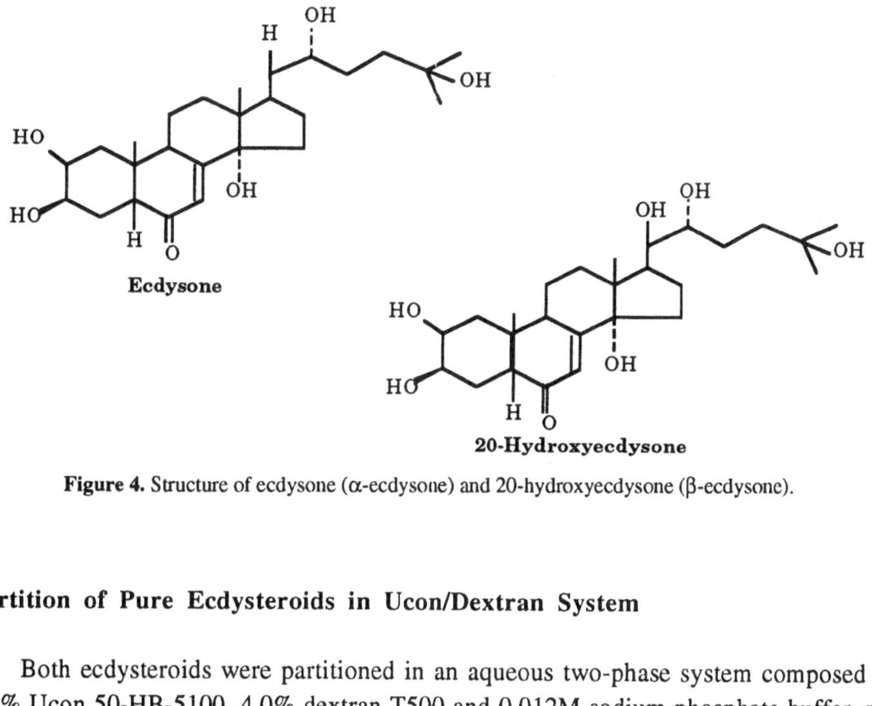

Figure 4. Structure of ecdysone (α-ecdysone) and 20-hydroxyecdysone (β-ecdysone).

Partition of Pure Ecdysteroids in Ucon/Dextran System

Both ecdysteroids were partitioned in an aqueous two-phase system composed of 5.0% Ucon 50-HB-5100, 4.0% dextran T500 and 0.012M sodium phosphate buffer, pH 7.0.[10] The K value, which reflects the relative affinity of a compound for the two phases, is in part a reflection of hydrophobicity. Since Ucon is more hydrophobic than dextran, different partitioning behavior can be expected for the two ecdysteroids, depending on their hydrophobicity. This trend was not reflected in their partitioning in the primary Ucon/dextran system, with K = 1.12 for ecdysone and K = 1.30 for 20-hydroxyecdysone.

However, when upper Ucon phase was removed and the temperature increased to 56°C, the hydrophobicity of ecdysone resulted in its partitioning more to lower Ucon phase (K = 0.59) than the more hydrophilic 20-hydroxyecdysone (K = 1.34).[10]

The addition of ethanol influenced ecdysteroid partition. Since Ucon is more hydrophobic than dextran, ethanol can be expected to partition preferentially to the upper Ucon-rich phase, thereby increasing the hydrophobicity of this phase. Indeed, K for both ecdysteroids increased as the concentration of ethanol increased. The best results were obtained in the system with 20% ethanol (Table 5).

Table 5. Partition of α-ecdysone and β-ecdysone (20-hydroxyecdysone). Primary phase systems: 6.0% Ucon 50-HB-5100, 5.0% dextran T500, 1.0 mg α-ecdysone, 0.5 mg β-ecdysone, 20.0% ethanol and 0.012 M sodium phosphate bufffer, pH 7.0; 0.1 M sodium chloride added to aqueous Ucon phase prior to increasing temperature. K and G values at 56° C are for partition between water and Ucon phases formed by increasing the temperature. Y = % yield of ecdysones in the water phase.

	K(22°C)	G(22°C)[1]	K(56°C)	G(56°C)[2]	Y
α-Ecdysone	3.01	12.02	0.62	3.94	73.6
β-Ecdysone	3.18	12.72	1.91	12.09	85.6

[1]Volume ratio (V_t / V_b) at 22°C = 4.0.

[2]Volume ratio (V_t / V_b) at 56°C = 6.3.

Purification of Ecdysone and 20-Hydroxyecdysone from Spinach

The results obtained in partitioning these ecdysteroids suggest that it is possible to purify them using temperature-induced phase separation, with a purification scheme similar to the one proposed in Figure 2. Based on the results presented previously it can be expected that when a cell homogenate containing ecdysteroids is added to a Ucon/dextran or Ucon/hydroxypropyl starch system, the ecdysteroids will partition to the upper Ucon-rich phase, while most protein and other cell debris will be enriched in the lower dextran or HPS phase. The lower dextran or HPS phase would go to waste. The upper Ucon phase would be subjected to temperature-induced phase formation, resulting in formation of a new two-phase system as previously described. In this system, the ecdysteroids would be obtained purified from contaminating compounds in the final upper water phase. The lower Ucon phase is recovered and can be reused in further extractions.

To test this purification process, spinach leaves, which contain significant amounts of both ecdysone and 20-hydroxyecdysone, were homogenized and sonicated for 15 min in a solution comprised of 100 g Ucon 50-HB-5100 and 294 g phosphate buffer (0.1M, pH = 7.0).[11] Extraction was allowed to proceed for 72 hours at 4.0°C. Extract was then centrifuged at 12,000 rpm for 20 min to remove plant particulates. Final extract had a Ucon concentration of 25.4%.

The Ucon-spinach extract was mixed with an aqueous stock solution of Reppal PES 200 (21% w/w) to a final concentration of 11.0% Ucon, 7.5% Reppal. Systems were made both with and without added ethanol. Phase systems were separated by centrifugation at 125g for 10 min, after which upper Ucon phase was removed. Both ecdysteroids partitioned strongly to upper Ucon phase in this step (Table 6). Containers with Ucon phase were placed in a water bath at 56°C for 20 min to allow formation of a new aqueous two-phase system, consisting of a Ucon-rich lower phase and a water/buffer upper phase. Total yield was calculated using the upper water phase at 56°C (Table 6). Due to the large volume of water phase, the yield of both steroids was high in this phase.[11]

Table 6. Partition of α-Ecdysone and β-Ecdysone from Spinach. System 1: 11.0% Ucon 50-HB-5100, 7.5% Reppal PES 200 and 0.1M sodium phosphate buffer. System 2: Same as System 1, with 20% ethanol content.

Compound	System	K (22°C)	G (22°C)	K (56°C)	G (56°C)	Y
α-Ecdysone	1	50.3	159	0.51	4.28	80.5
	2	>100	>100	1.74	7.0	88.7
β-Ecdysone	1	30.6	94.5	2.16	18.1	93.8
	2	>100	>100	1.94	11.7	91.2

CONCLUSIONS

A new purification technique for biomolecules has been developed by combination of temperature-induced phase separation and partitioning in aqueous two-phase systems. The technique is based on the use of a thermo-separating polymer as one of the phase-forming polymers in an aqueous two-phase system. The thermo-separating polymer can be removed from the protein by a moderate temperature increase. All proteins so far studied have been excluded from the polymer phase formed above the cloud point. The target protein is obtained in a water/buffer solution after only two purification steps. A polymer with attached affinity ligand can be separated from the protein solution by raising the temperature above the cloud point of the polymer. The temperature of phase separation can be controlled by the salt concentration and by the polymer hydrophobicity. The ethylene oxide-propylene oxide random copolymer Ucon 50-HB-5100, with a cloud point of 50°C, was successfully used. With this polymer it was possible to use a covalently bound affinity ligand for enzyme purification. The use of affinity ligand increased the specificity of the purification technique.

An ethylene oxide-propylene oxide random copolymer with a cloud point of 18°C has been synthesized and used for protein purification. With polymers which have low cloud point it is possible to perform temperature-induced phase separation at temperatures close to room temperature. With the use of a hydrophobic polymer with low cloud point relatively high purification factors were achieved. This was due to the partitioning of hydrophilic proteins to the lower phase in the primary phase system. More hydrophobic proteins could be partitioned to the phase containing the hydrophobic thermo-separating polymer. It was thus possible to purify proteins by utilizing protein surface hydrophobicity by application of the combination of aqueous two-phase partitioning and temperature-induced phase separation. The purification process could also be used for the purification of steroids from a crude homogenate.

Temperature-induced phase separation makes it possible to achieve important simplifications when aqueous two-phase systems are used in bioseparations. The polymer can easily be removed from the target biomolecule. The polymer can be reused for repeated extractions. The purified target biomolecule is obtained in a water solution which facilitates integration with other purification techniques.

Acknowledgements

This research is supported by grants from the Swedish National Board for Industrial and Technical Development (NUTEK) and from the Swedish Research Council for Engineering Sciences (TFR). Berol Nobel AB, Stenungsund, Sweden, is thanked for the synthesis of the 20% ethylene oxide/80% propylene oxide random copolymer.

REFERENCES

1. S. Saeki, N. Kuwahara, M. Nakata and M. Kaneko, Upper and lower critical temperatures in poly(ethylene glycol) solutions, *Polymer*, 17:685 (1976).
2. H.-O. Johansson, G. Karlström and F. Tjerneld, Experimental and theoretical study of phase separation in aqueous solutions of clouding polymers and carboxylic acids, *Macromolecules*, 26:4478 (1993).
3. C. Bordier, Phase separation of integral membrane proteins in Triton X-114 solution, *J. Biol. Chem.*, 256:1604 (1981).
4. P.-Å. Albertsson, "Partition of Cell Particles and Macromolecules," 3rd ed., Wiley, New York (1986).
5. H. Walter, D.E. Brooks and D. Fisher, "Partitioning in Aqueous Two-Phase Systems," Academic Press, Orlando (1985).
6. F. Tjerneld, Aqueous two-phase partitioning on an industrial scale, *in*: "Poly(Ethylene Glycol) Chemistry: Biotechnical and Biomedical Applications," J.M. Harris, ed., Plenum Press, New York (1992).

7. P.A. Harris, G. Karlström, and F. Tjerneld, Enzyme purification using temperature induced phase formation, *Bioseparation*, 2:237 (1991).
8. P.A. Alred, A. Kozlowski, J.M. Harris, F. Tjerneld, Application of temperature-induced phase separation at ambient temperature for enzyme purification, *J. Chromatogr.*, 659:289 (1994).
9. P.A. Alred, F. Tjerneld, A. Kozlowski, J.M. Harris, Synthesis of dye conjugates of ethylene oxide-propylene oxide copolymers and application in temperature-induced phase partitioning, *Bioseparation*, 2:363 (1992).
10. P.A. Alred, F. Tjerneld and R.F. Modlin, Partition of ecdysteroids using temperature-induced phase separation, *J. Chromatogr.*, 628:205 (1993).
11. R.F. Modlin, P.A. Alred and F. Tjerneld, Utilization of temperature-induced phase separation for the purification of ecdysone and 20-hydroxyecdysone from spinach, *J. Chromatogr.*, 668:229 (1994).
12. G. Kopperschläger, Affinity extraction with dye ligands, *in*: "Methods in Enzymology, Vol. 228, Aqueous Two-Phase Systems," H. Walter and G. Johansson, eds., Academic Press, San Diego (1994).

POLY(ETYHLENE GLYCOL)-PROTEIN INTERACTION IN SALT CONTAINING AQUEOUS SOLUTIONS

Andres Veide,[1] Cynthia Hassinen,[1] Dan Hallén,[2] Mark Eiteman,[3] Bo Lassen,[4] and Krister Holmberg[4]

[1]Department of Biochemistry and Biotechnology
Royal Institute of Technology, S-100 44 Stockholm, Sweden
[2]Thermochemistry, Chemical Centre
Lund University, S-221 00 Lund
[3]Department of Biological and Agricultural Engineering
University of Georgia, Athens, GA 30602, USA
[4]Institute for Surface Chemistry
P.O. Box 5607, S-114 86 Stockholm, Sweden

INTRODUCTION

Poly(ethylene glycol) (PEG) is a common polymer in aqueous two-phase systems, and it is also used in other fields of biology, e.g. (i) protein precipitation, (ii) for creation of biocompatible surfaces, (iii) for *in vitro* stabilization of proteins, and (iv) as a refolding enhancer. Although a general picture of the features governing protein partition in PEG/salt aqueous two-phase systems can be given, the molecular mechanisms behind it are poorly understood. There is need for partition experiments performed with well characterized model peptides and proteins. We have developed a concept where we are using genetic engineering to modify a staphylococcal protein A derivative, ZZ0, with respect to its content of different short peptide units, e.g. like AlaTrpTrpPro and AlaIleIlePro. We are particularly interested in the use of these protein derivatives and free short peptides for the study of the molecular mechanisms involved in the PEG-protein interaction. Several techniques can be used to probe PEG-free peptide and PEG-protein interactions, not only in bulk solution, but also at solid surfaces. In this paper we will describe results from partition in PEG/potassium phosphate aqueous two-phase systems, and interaction with PEG coated surfaces. For the latter measurements we have applied: (i) retention studies on packed bed chromatography columns, and (ii) ellipsometry.

We are also setting up a system where the thermodynamics of the partitioning process will be characterized using microcalorimetric technique. This could prove to be an useful complement to the ongoing theoretical modelling of the free energy of the partitioning in aqueous two-phase systems.

Aqueous Biphasic Separations: Biomolecules to Metal Ions
Edited by R.D. Rogers and M.A. Eiteman, Plenum Press, New York, 1995

133

MODEL COMPOUNDS

The model peptides and proteins are schematically shown in Figure 1. The peptides have been prepared with a peptide synthesizer (Advanced Chemtech, Louisville, KY, USA) at University of Georgia. Z is a hydrophilic synthetic IgG binding domain derived from the B domain of staphylococcal protein A (SpA)[1] having a pI of about 4.8.[2] The three-dimensional solution structure of the recombinant B domain of SpA is composed of a bundle of α-helices with most of the hydrophobic residues buried in the interior of the bundle.[3] The conformation of the C-terminal peptide stretch, into which the different peptide units are cloned, is not known. The design, cloning, cultivation of plasmid harbouring *Escherichia coli* (*E. coli*) RV308, and purification of the ZZ based protein derivatives have been described earlier.[4,5]

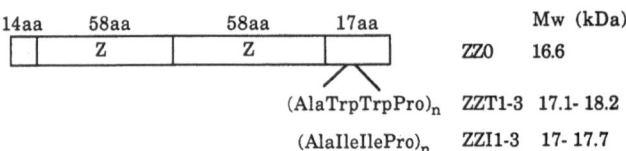

Figure 1. ZZ0 based model proteins. Z is a synthetic IgG-binding domain derived from staphylococcal protein A. (From C. Hassinen, K. Köhler, and A. Veide, J. Chromatogr. A 668:121 (1994), Elsevier Science, with permission.)

AQUEOUS TWO-PHASE SYSTEMS AND HYDROPHOBIC INTERACTION CHROMATOGRAPHY

The idea to begin with short Trp rich peptide units in the interaction study evolved from partitioning research with *E. coli* ß-galactosidase,[6] and interpretation of literature data on dipeptide partition.[7] ß-galactosidase contains an unusually high amount of Trp residues,[8] and partitions strongly to the PEG phase in PEG 4000/potassium phosphate aqueous two-phase systems.[6] The non polar amino acid Ile was selected as a control to Trp.

The difference between the Trp and Ile residues on effecting the partitioning of proteins in PEG/potassium phosphate aqueous two-phase systems has been demonstrated.[4,5] Partition coefficients of various short peptides and modifications of ZZT0 in PEG 8000/potassium phosphate systems are shown in Figure 2. It clearly illustrates that molecules enriched in Trp residues partitions much stronger to the top PEG-rich phase compared to Ile residues.

If PEG is immobilized on a solid surface we observe a retention behaviour which qualitatively is related to the behaviour in the bulk system (Figure 3).[5] A more detailed study of the surface interaction will demand a well characterized PEG surface (see below).

The difference in partitioning/retention behaviour between Ile and Trp can be understood from their different structures. The side chain of Ile is a nonpolar aliphatic structure while the side chain of Trp is much more polar and can become involved in hydrogen bond formation and charge transfer interaction. It is clear that the two molecules have different properties, and those determine their respective interaction with other molecules.

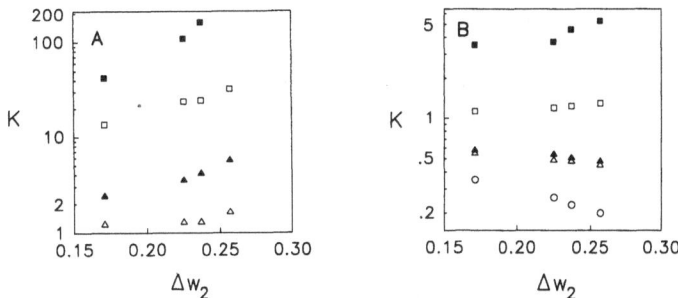

Figure 2. Partition coefficients of (A) different peptides: AlaIleIlePro (Δ), AlaTrpTrpPro (□), (AlaIleIlePro)$_2$ (▲), (AlaTrpTrpPro)$_2$ (■), and (B) ZZ protein derivatives: ZZ0 (○), ZZI1 (Δ), ZZT1 (□), ZZI2 (▲), ZZT2 (■) in PEG 8000/potassium phosphate (mole ratio base/acid = 0.11; pH ≈ 5.5) systems at 25°C versus the difference in PEG concentration between the phases (Δw$_2$). Partition coefficient, K, is the top to bottom phase concentration ratio of partitioned molecules. In this PEG/potassium phosphate system the top phase is PEG-rich and the bottom phase is potassium phosphate-rich. (From M. A. Eiteman, C. Hassinen, and A. Veide, Biotechnol. Prog., in press (1994), American Chemical Society, with permission.)

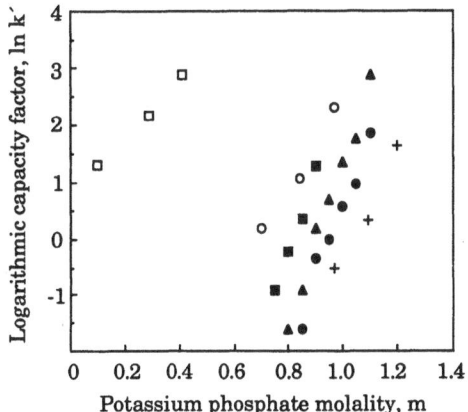

Figure 3. Hydrophobic interaction chromatography of ZZT$_n$ and ZZI$_n$ proteins on PEG (Hydropore-HIC Dynamax-TI bed from Rainin Instrument Co. Inc., U.S.A.). ZZ0 (+); ZZT1 (○); ZZT3 (□); ZZI1 (●); ZZI2 (▲); ZZI3 (■). The retention, expressed as the logarithmic capacity factor (lnk′), is plotted as a function of potassium phosphate concentration (dibasic/monobasic phosphate mole ratio = 1.42). The capacity factor is defined as k′ = (V$_r$/V$_0$)-1, where V$_r$ and V$_0$ are the retention and void volumes, respectively. (From C. Hassinen, K. Köhler, and A. Veide, J. Chromatogr. A 668:121 (1994), Elsevier Science, with permission.)

PEG

Protein stabilizer

To get a picture of how PEG interacts with proteins let us shortly consider the role PEG and, for comparison, phosphate ions play as protein stabilizers.[9] Protein stabilizers seem to obey a general rule: co-solvents which, at high concentration, stabilize the native structure of proteins are preferentially excluded from the surface of a protein.

On the on hand, phosphate ions (especially dibasic and tribasic), which are excluded from the protein surface through a mechanism referred to as the surface tension effect, are stabilizing folded protein structures at different types of physico-chemical conditions (e.g. elevated temperatures, extreme pH). The stabilizing properties of salt ions (and salting-out effectiveness) increase with their effectiveness in increasing the surface tension of water.

On the other hand, the interaction of PEG with proteins will depend to a great extent on the surrounding conditions. At stabilizing conditions proteins are sterically excluded from PEG. However, at denaturing conditions, when a protein structure gets more flexible or starts to unfold, PEG, due to its hydrophobic character, can bind to unfolded proteins and folding intermediates. It has been described that only a certain folding intermediate of bovine carbonic anhydrase B binds to PEG.[10] This property of PEG has been explored for the enhancement of protein refolding.[11]

Interaction with Different Molecules

The formation of the PEG/phosphate aqueous two-phase system can be seen as salting-out of PEG by phosphate, which is related to the surface tension effect of phosphate on water. The interaction of PEG with salt ions is important to understand, and electrophoretic mobility measurements on PEG demonstrate that some anions, e.g. iodide, in contrast to phosphate, seem to bind to PEG and make it act as carrying negative charge.[12]

PEG is described as a strong hydrogen bond acceptor due to the paired ether oxygen electrons, which provides for its water solubility.[13] It can form complexes with monomeric and polymeric electron acceptors (e.g., hydrogen bond donors).[14] A specific interaction between PEG and a protein UV_{280}-chromophore has been suggested to occur as an explanation for large perturbations on the UV spectra of proteins in PEG solutions.[15] The chromophore could very well be that of Trp. It has been shown that the water solubility of Trp is increased by the addition of PEG.[16]

When we studied the Trp fluorescence spectra of proteins ZZT1 and ZZT2 in potassium phosphate aqueous solutions where the PEG 4000 concentration was increased from 0 to as much as 20 % (w/w) no shift in the spectra was detected (Figure 4). Although this indicated that the polarity around the Trp-residues was not shifted with the addition of PEG 4000 at different salt concentrations, there still is a possible presence of a weak interaction between PEG and Trp.

Another important observation is that the hydrophobic character of PEG (MW range 100-4000) is not pronounced in 20 % aqueous solutions at 25°C. This was demonstrated with binding study experiments between PEG and ANS (1-anilino-8-sulfonate).[16]

Altogether, the existence of a weak attractive force between PEG and protein can not be excluded. The structural properties of the protein (stability and flexibility) should be important for this interaction, and they will be affected by the aqueous solutes creating the chemical and physical environment of the protein molecule.

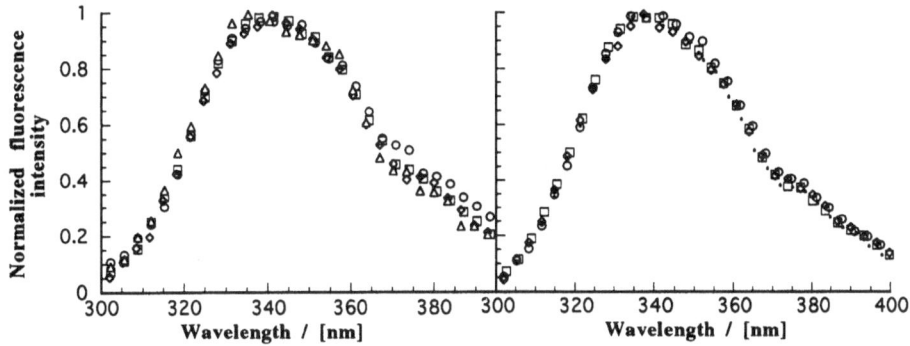

Figure 4. Trp-fluorescence from ZZT1 (left) and ZZT2 (right) in different PEG 4000 - potassium phosphate solutions. % (w/w) PEG 4000/molality potassium phosphate: (O) 0/0.44, (□) 1/0.9, (◇) 10/0.44, (Δ) 20/0.35. Excitation wavelength was 295 nm.

MICROCALORIMETRIC MEASUREMENTS

To get a more complete description of the partitioning of peptides and proteins in aqueous two-phase systems it would be useful to have the full thermodynamic characteristics of the systems. Microcalorimetry is used in many areas which are of interest for biochemistry and biology such as studies on living cells (in suspension or mammalian tissues), protein ligand binding studies, solution properties of biochemical model compounds and stability control of different materials. The non-specificity of the technique is often an advantage, since the calorimeter detects all processes that occur in the system which has an enthalpy different from zero. However, calorimetry has not yet been used in any study of partition of proteins or other crystalline compounds in aqueous polymer phases, mainly due to the lack of appropriate microcalorimetric methods to measure enthalpies of solution of small amounts of solids.

Recently a new solution microcalorimeter for solids, which needs relatively small quantities of material, was developed by the Thermochemistry Department at Lund University.[17] From microcalorimetric measurements with this new technique one can obtain partition coefficients, the temperature dependence of the partition coefficient and the total difference in enthalpy for the solute between the two phases. Through the thermodynamic measurements we can probe how the small alterations in protein sequence, created by the insertion of the short peptide units into the ZZ protein, can be related to changes in interaction with different molecules. (e.g., PEG or salt).

All theoretical work is at the moment based on modelling of ΔG for the solute in the two phases. In the mean-field lattice theory there are parameters describing the interactions between the different components in the mixture, the χ-parameters. Ideally these parameters are enthalpic interaction parameters. However, this is not the case for aqueous polymer mixtures, where these parameters contain entropic contributions, due to the small size of the water molecule and special properties of water. Thus, experimentally found enthalpies for these systems can not directly be incorporated in the mean-field lattice model as it is defined at the moment. Comparison of the experimentally found enthalpy of transfer, ΔH_p, and the difference in calculated χ-parameter, $\Delta \chi_p$, for the protein will give us information about the entropic contribution of the χ-parameter.

Initial measurements of the enthalpies of transfer between different aqueous solutions for the ZZ0 and ZZT1 model proteins have been performed showing that the experimental system is working (Table 1). One can observe the relatively large but opposite effect on the addition of potassium phosphate to a PEG 4000 solution has on ΔH_t for the two proteins. This indicates the important role salts play in the partitioning behaviour of proteins. Although ΔH_t was increased for the ZZ protein in a salt environment by the insertion of AlaTrpTrpPro (Table 1, column 2), ΔH_t for ZZ actually decreased a little in a salt free environment for the same peptide insertion (Table 1, column 1).

Table 1. Initial measurements of enthalpies of transfer, ΔH_t (J/g), for ZZ0 and ZZT1.

Protein	Water to 20% (w/w) PEG 4000 in water	Water to 20% (w/w) PEG 4000 in 4% (w/w) potassium phosphate (pH7)
ZZ0	38.9	4.9
ZZT1	33.3	67.4

ELLIPSOMETRY

By creating a well characterized PEG surface we would like to compare the interaction in bulk solution with the interaction at a solid surface. The PEG immobilization would allow a study of the PEG-protein interaction independent of the salt concentration. Various methods,

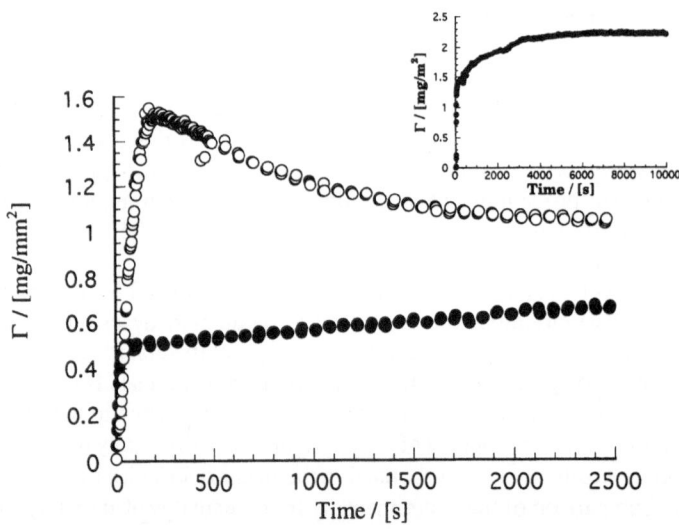

Figure 5. Ellipsometric measurements of adsorption of Pluronic F 127 (insert right corner) (500ppm) on HMDSO modified Si wafer, ZZ0 (●) (25 ppm) on Pluronic F 127 in 0.5 M potassium phosphate, pH 7, and ZZT2 (O) (25 ppm) on Pluronic F 127 in 0.5 M potassium phosphate, pH 7.

such as ellipsometry, exists to create controlled PEG surfaces on which interaction measurements can be made with different techniques. Preliminary ellipsometric studies were performed in a system with a hydrophobically coated base matrix on to which a PEG-PPO [poly(propylene oxide)] -PEG block copolymer (Pluronic F127) had been adsorbed. The PPO part of the molecule is attaching the molecule to the surface leaving the two PEG end parts of the block copolymer to solubilize in the aqueous environment. Results from initial adsorption experiments with model proteins ZZ0 and ZZT2 onto this surface measured by ellipsometry are described in Figure 5. It is illustrated how the PEG-PPO-PEG block copolymer is being adsorbed to the surface, and then how both ZZ0 and ZZT2 are absorbed to the precoated surface. As expected from partitioning and HIC experiments more of the ZZT2 is adsorbed than ZZ0.

Since the time stability of this surface is inadequate, we are trying another immobilization technique, depending on strong adsorption of positively charged PEG-poly(ethylene imine) (PEI) copolymer to a negatively charged surface coating,[18] which so far looks promising.

·CONCLUDING REMARKS

We have shown that large effects in PEG containing systems (aqueous two-phase systems, chromatography support) can be caused by small specific changes in the protein amino acid sequence. PEG has a widespread use in the field of biology and this should emphasize research aimed at the understanding of the PEG-protein molecular interactions. This work points out our efforts in this direction where the combination of different techniques, bulk as well as surface based ones, is particularly employed. As we learn more about the molecular mechanisms underlaying the interactions the use of phase systems for characterization of protein surfaces can become one new interesting application. In protein purification processes the selectivity could be enhanced in a more controlled manner by the addition of properly designed peptide sequences to the target protein.

ACKNOWLEDGMENTS

This work was supported by grants from Teknikvetenskapliga Forskningsrådet (Swedish Research Council for Engineering Sciences) and Helge Ax:son Johnsons Stiftelse, Sweden. Dr. John Wunderlich is acknowledged for synthesis of peptides.

REFERENCES

1. B. Nilsson, T. Moks, B. Jansson, L. Abrahmsén, A. Elmblad, E. Holmgren, C. Henrichson, C., T.A. Jones, and M. Uhlén, A synthetic IgG-binding domain based on the staphylococcal protein A, *Protein Engineering*, 1:107 (1987).
2. C. Ljungquist, A. Breitholtz, H. Brink-Nilsson, T. Moks, M. Uhlén, and B. Nilsson, Immobilization and affinity purification of recombinant proteins using histidine peptide fusions, *Eur. J. Biochem.*, 186:563 (1989).
3. H. Gouda, H. Torigoe, A. Saito, M. Sato, Y. Arata, and I. Shimada, Three dimensional solution structure of the B domain of staphylococcal protein A, *Biochem.*, 31:9665 (1992).
4. K. Köhler, C. Ljungquist, A. Kondo, A. Veide, and B. Nilsson, Engineering proteins to enhance their partition coefficients in aqueous two-phase systems, *Bio/Technology*, 9:642 (1991).

5. C. Hassinen, K.Köhler, and A. Veide, Poly(ethylene glycol) - potassium phosphate aqueous two-phase systems: Insertion of short peptide units into a protein and the effects on partitioning, *J.Chromatography A*, 668:121 (1994).

6. A. Veide, "Aqueous Two-Phase Partitioning: A Technique for Large Scale Purification of Microbial Proteins," PhD Thesis, Royal Institute of Technology, Stockholm, ISBN 91-7170-910-X (1987).

7. A.D. Diamond, X. Lei, and J.T. Hsu, Reversing the amino acid sequence of a dipeptide changes its partition in an aqueous two-phase system, *Biotechnol. Technol.*, 3:271 (1989).

8. A. Kalnins, K. Otto, U. Rüther, and B. Müller-Hill, Sequence of the *lac Z* gene of *Escherichia coli*, EMBO J., 2:593 (1983).

9. S.N. Timasheff and T. Arakawa, Stabilization of protein structure by solvents, *in*: "Protein Structure: a Practical Approach," T.E. Creighton, ed. ,IRL Press, Oxford (1990).

10. J.L. Cleland and T.W. Randolph, Mechanism of poly(ethyleneglycol) interaction with the molten globule folding intermediate of bovine carbonic anhydrase B, *J. Biol. Chem.*, 267:3147 (1992).

11. J.L. Cleland, S.E. Builder, J.R. Swartz, M. Winkler, J.Y. Chang and D.I.C. Wang,Poly(ethyleneglycol) enhanced protein refolding, *Bio/Technology*, 10:1013 (1992).

12. G. Johansson, Effects of different ions on the partition of proteins in aqueous dextran-poly(ethylene glycol) two-phase systems, *in*: "Proceedings of the International Solvent Extraction Conference 1971," Society of Chemical Industry, London (1971).

13. F.E. Bailey Jr. and J.V. Koleske, "Poly(ethylene oxide),"Academic Press, New York (1976).

14. F.W. Stone and J.J. Stratta, 1,2-epoxide polymers, *in*: "Encyklopedia of Polymer Science and Technology," H. F. Mark, G. Gaylord, and N.M. Biales, ed., Wiley Interscience, New York (1966).

15. M. Laskowski Jr., Measurment of accessibility of protein chromophores by solvent perturbation of their ultraviolet spectra, *Federation Proc.*, 25:20 (1966).

16. K.C. Ingham, Poly(ethyleneglycol) in aqueous solution: Solvent perturbation and gel filtration studies, *Arch. Biochem. Biophys.*, 184:59 (1977).

17. D. Hallén, E. Qvarnström, and I. Wadsö, A multiple sample microcalorimeter for dissolving solids. In manuscript.

18. C. Brink, E. Österberg, K. Holmberg, and F. Tiberg, A method to obtain dense grafting of poly(ethylene glycol) or polysaccharide to polystyrene, *Colloids Surfaces*, 66:149 (1992).

INTERFACIAL EVENTS IN PHASE SEPARATION AND

CELL PARTITIONING IN AQUEOUS TWO-PHASE SYSTEMS

Frank D. Raymond[1] and Derek Fisher[2,3]

[1]Department of Metabolic Medicine
Royal Postgraduate Medical School
Hammersmith Hospital
DuCane Road
London W12 ONN, UK

[2]Molecular Cell Pathology
Royal Free Hospital School of Medicine
Rowland Hill Street
London NW3 2PF, UK

SUMMARY

The processes by which phase separation occurs in aqueous two-phase systems of polyethylene glycol and dextran are described in terms of the fluid dynamics of phase droplets and phase streams. Interactions of cells with the surfaces of these microphases early in phase separations determine the cell partition coefficient, which is conventionally measured when phase separation is virtually completed.

INTRODUCTION

In the technique of cell partitioning, cells are added to buffered solutions of polyethylene glycol (PEG) and dextran which separate into two immiscible phases above certain concentrations of the polymers.[1-4] Cells distribute between the two phases and the horizontal, bulk interface on the basis of surface properties. In phase systems that possess an electrostatic potential difference between the phases ($\Delta\psi$) due to the uneven partitioning of ions, membrane surface charge is important in influencing cell partitioning ("charge-sensitive" partitioning), whereas in systems of low $\Delta\psi$ partitioning has been considered to be determined by non-charged membrane components, such as membrane lipids ("non-charge sensitive" partitioning).[5] Our early studies on cell partitioning in phases of low $\Delta\psi$ indicated dynamic features that had not previously received attention. [6-8] This led us to propose that cell partitioning depends closely upon the mechanism by which phase systems separate, and the interactions of cells with the surfaces of the complex microphases that constitute the phase separation process. We have drawn attention to the implications these dynamic features have for obtaining cell separations in aqueous two-phase systems. [9-10]

Cell partitioning is usually measured, and the transfer step made in counter current

---

[3] To whom correspondence should be addressed

Aqueous Biphasic Separations: Biomolecules to Metal Ions
Edited by R.D. Rogers and M.A. Eiteman, Plenum Press, New York, 1995

141

distribution (CCD), when phase separation is close to completion. The associations of cells with the phase droplets that represent these late stages of phase separation correlate with the partition behaviour. [11] However, as we outline in this article, it is during the early events in phase separation that cell-phase interface interactions occur that determine the late events, and consequently the cell partitioning.

METHODS

Dextran T500 was obtained from Pharmacia Fine Chemicals, Uppsala, Sweden and PEG 6000 from BDH Ltd, Poole, UK. Stock solutions were prepared by weight at 20° in 0.15 NaCl buffered with sodium phosphate, pH 6.8 from an approximately 20% w/w solution of dextran, standardised by polarimetry on the assumption of a specific rotation of 199°, and solid PEG. Systems were centrifuged at 2000g x 60 min in an MSE Mistral 6L centrifuge at 20°: the top and bottom phases were removed by aspiration and stored at 4°. For experiments phase systems were reconstituted with 1g of each phase. Phase compositions of polymers are given in abbreviated form: 5D-5P represents 5% w/w dextran and 5% w/w PEG. As the properties of phase systems are dependent on the molecular weight distributions of polymers used, particularly in systems close to the critical point, batch numbers are given with individual experiments. Cell partition and microscopic examinations were performed as described previously.[6]

RESULTS

General features of phase separation

From observations of a large number of phase systems at varying distances from the critical point, we found that the process of phase separation involves 4 stages, the duration of which depends critically on the phase composition.

1) **The immobile stage**. This stage occurred directly after mixing and continued until bulk movement of the phases (stage 2) began. After mixing, the system generally appeared turbid due to the presence of a network of interfaces. Turbidity increased with increasing distance from the critical point, and systems very close to the critical point were initially (and briefly) clear. The network of interfaces separated dextran-rich areas from PEG-rich areas, giving the system a pitted appearance. The pitted areas increased in size by coalescence to form domains of PEG-rich phase and dextran-rich phase. Further coalescence led to the formation of streams of adjacent PEG-rich phase and dextran-rich phase which, on reaching a critical size, began to move upward and downwards respectively. The time taken for motion to be observed ($T_{immobile}$) decreased with increasing polymer concentration. As we show later, it is at this immobile stage that cell partition is determined in phase systems containing cells.

2) **Bulk phase separation stage**. In this stage bulk separation of the phase system began and was completed, and the most obvious differences in the pattern of phase separation between systems of increasing polymer concentration were observed. For the polymers dextran T500 (Batch no 2836) and PEG 6000 (lot number 6800790) systems close to the critical point (e.g., 5D-3.45P; 5D-3.5P; 5D-3.75P; 5D-4P) cleared by a slow, gradual contraction of diffuse streams of phase, whereas phases much further away from the critical point (e.g., 5D- 4.75; 5D-5P; 5D-5.5; 5D-6P) separated by a comparatively rapid contraction of turbidity into a band with relatively clear phases developing above and below. The speed of contraction of the band was in the order:
5D-5.5P > 5D-5P > 5D-4.75P > 5D-6P.
In systems of intermediate composition (e.g., 5D-4.25; 5D-4.5P) both processes operated. As we describe later, it is generally within this stage that the best opportunity for separating different cells occurs. The widely held practice of using a "usual sampling time", without any optimisation of the sampling time for the particular separation being performed, may therefore miss the appropriate window for efficient separation.

3) **The microdroplet stage**. This stage is characterised by a dispersion of small droplets of about 5-10μ diameter, which cleared slowly from both phases to the horizontal interface, and was broadly similar in all phases studied. It is at this stage that cell partitions are most generally measured in single tube partition, or the transfer step made in CCD.

4) **Equilibrium**. At this stage both phases are free of phase droplets. Equilibrium can be rapidly achieved by centrifugation and is the stage at which the partition of soluble substances such as proteins is usually measured.

Microscopic examination of phase separation

Microscopic examination of phase systems as they separated showed that the patterns of phase separation depended on the rate of coalescence of PEG-rich areas and the coalescence of dextran-rich areas at the start of phase separation. Two types of coalescence mechanism were observed, one predominating in phases close to the critical point and the other in phases far from the critical point (Figure 1). In the former, coalescence progressed slowly, and the film of the opposite phase solution that was present between two coalescing phase domains as they approached each other had sufficient time to drain away before coalescence occurred. By contrast, in the latter systems fast coalescence occurred and the intervening film appeared to have insufficient time to drain away before coalescence occurred, resulting in the trapping of droplets of one phase in the other, thereby forming multiple phase droplets.

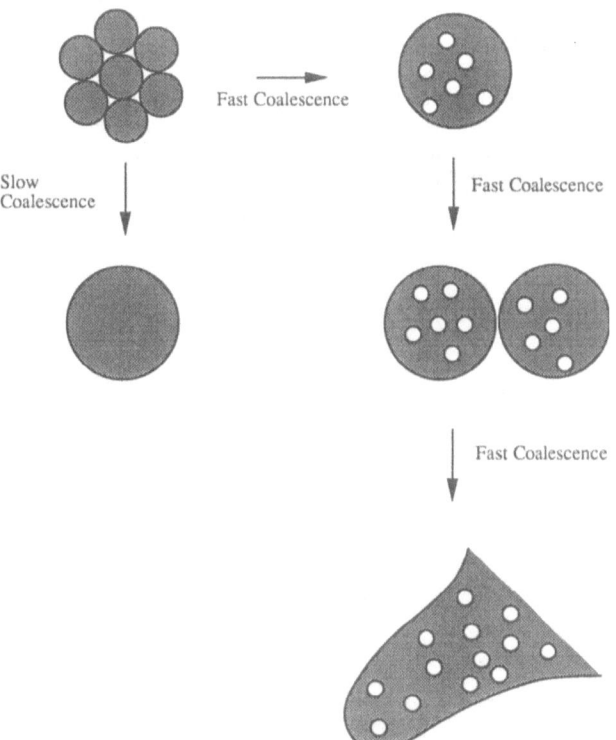

Figure 1. Fast and slow coalescence mechanisms. Dextran-rich phase droplets (dark) in PEG-dextran phase systems close to the critical point coalesce slowly to form larger droplets. In phases further from the critical point coalescence is fast, leading to entrapment of PEG-rich phase (clear).

It was also clear that phase separation showed distinct differences in fluid dynamics that were dependent on the polymer concentrations. Phase system of 5D-4P, 5D-4.25P, 5D-4.5P, 5D-5P, 5D-5.5P (Dextran T500, lot no 5287, and PEG 6000 lot no 681668) of low $\Delta\psi$ were examined. (It should be noted that with these polymer batches the 5D-3.5P systems did not form two phases).

In the 5D-4P system the solution remained transparent for a few seconds after thorough mixing. A slight turbidity then developed. This transformed into pitted areas which expanded in size by coalescence until the system became mobile: streams of phase moved upwards and downwards, producing a swirling motion in the vertical plane. Droplets formed from these streams by fragmentation or budding, and being smaller than the streams remained longer in the top and bottom phases. The streams gradually contracted forming a band at the developing horizontal interface leaving a heterogeneous dispersion of droplets in the top and bottom phases. Droplets also showed coalescence to form larger droplets, which in the top phase moved faster downward. Progressive contraction of the streams and relatively slow movements of the droplets to the bulk interface and coalescence with it completed phase separation.

The 5D-4.25P system showed similar features to the 5D-4P but with the stages occurring more rapidly, and with stream droplets as an additional feature. Stream droplets are droplets flowing inside streams (Figure 2a) and were generally dextran rich-droplets flowing in PEG-rich streams, which were present in the bottom (dextran-rich) phase. Their addition to the top phase as the PEG-rich streams coalesced with the bulk interface provided another source of dextran-rich droplets in the top phase (Figure 2a). Multiple stream droplets (droplets within droplets within a stream; Figure 2b) were observed in the 5D-4.5P system. Although some coalescence of droplets occurred within the streams, it appeared that the rapid flow of the phases, upward and downward, diminished collision between droplets, and hence diminished coalescence. A feature of this system was that when PEG-rich streams coalesced with the bulk interface (Figure 2b), the dextran-rich stream droplets were catapulted into the top phase and carried well into the top phase, contributing greatly to the dextran-rich droplets in the top phase, formed by budding and fragmentation. Multiple stream droplets and coalescence of droplets within streams was more apparent in the 5D-5P system. In the 5D-5.5P systems budding and fragmentation were rarely observed whereas multiple stream droplets predominated. Because coalescence of stream droplets occurred readily, the stream droplets were generally larger than in the 5D-5.5P system. Multiple phase droplets occurred in both phases and by coalescing with the bulk interface were able to transfer their contents to both phases in a series of sequential coalescence (Figure 2b).

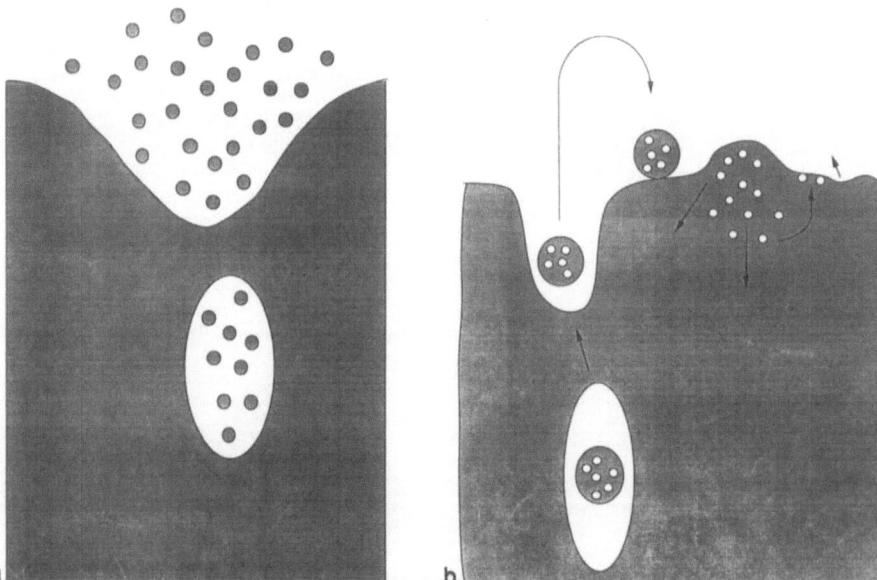

Figure 2. Stream droplets (a) and multiple stream droplets (b). Examples are shown of these in the dextran-rich phase (dark). By sequential coalescence with the bulk interface their contents are delivered into the bulk

In summary, as the phase systems move further away from the critical point (5D-4P --> 5D-5.5P) there is a progressive change in the relative contributions of the different mechanisms; phases move faster and coalescence is more rapid and leads to a sequential change in the pattern of the different phase structures.

Cell partitioning in phase system of low electrostatic potential difference

Phase systems of low $\Delta\psi$ used for cell partitioning have typically had compositions close to the critical point. [5] Phase separation was speeded up by laying the partition tube on its side, thereby providing a very short settling path. Under these conditions significant top phase partitions were obtained, that subsequently have been reported to have a relatively small time dependency.[10,11]

We have adopted an alternative approach to the use of these phase systems. A range of phase systems of increasing distance from the critical point has been used, and the distribution of cells between the developing top and bottom phases and the bulk interface has been measured at intervals as phase separation proceeded. Like Walter and his colleagues, we have used erythrocytes of different species as model cells with different surface properties.

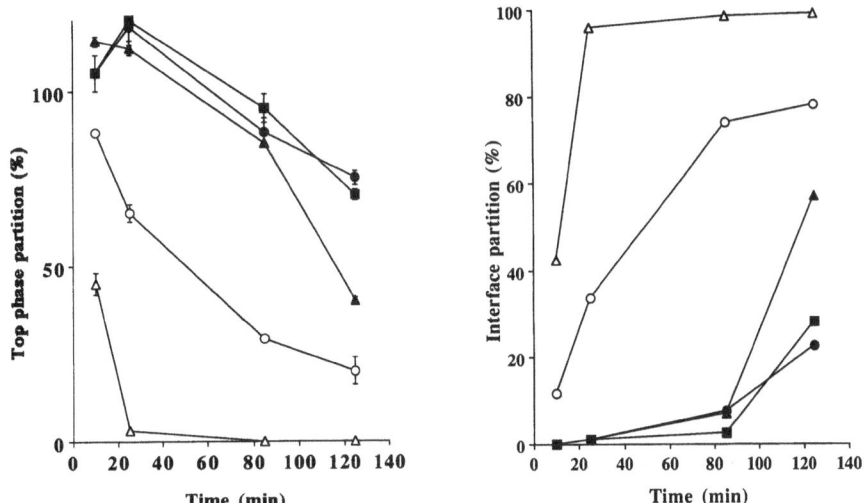

Figure 3. The time dependence of the top phase and interface partitioning of erythrocytes in 5D-4.05P low $\Delta\psi$ phase systems. The small bottom phase partition is not shown. (● rat, ■ turkey, ▲ chicken, ○ pig △ cattle).

Figure 3 shows the partition behaviour of erythrocytes of different species in a 5D-4.05P system (these were lots 4094 and 6255519) measured at different times. Partition is clearly a very time dependent process and the detection of different partitioning behaviour depended critically on the time of measurement. Rat, turkey and chicken cells had initially high top phase partitions that declined with time as cells partition to the interface. As a group, they partitioned very differently from pig and cattle erythrocytes, which showed a faster decline in top phase partition and faster accumulation at the interface. However, the rat, turkey and chicken cells partitioned very similarly and it was not until late sampling times (e.g, 125 min) that the chicken erythrocytes showed a markedly different (lower) partition than the rat and the turkey cells. Nevertheless, it is clear that by appropriate choice of partition times separation of cattle, pig and chicken cells could be distinguished (and therefore separated) from the other two.

On increasing the polymer concentration slightly, from 5D-4.05P to 5D-4.17P, marked alterations in the partition time courses were obtained (Figure 4). Rat, turkey, chicken and human erythrocytes show relatively high initial top phase partitions which declined with time as cells accumulated at the horizontal interface. By contrast, pig and cattle erythrocytes showed low top phase partition, initially appearing in the bottom phase but accumulating at the interface with time. In this phase system a very clear difference in partitioning between rat, turkey and chicken cells was obtained. The clear resolution of cattle and pig cells in the top phase seen with the 5D-4.05P system was lost, as top phase partition was low, but markedly different partition in the bottom phase and at the interface were obtained. Only the cattle and pig cells gave significant bottom phase partition. However, increasing the polymer concentration further, to 5D-5P, shifted the partition of all these cells to time dependent partitioning between the bottom phase and the interface.

Taken together, these data indicate that the relative affinity of these cells for the top, PEG-rich phase is:

rat > turkey > chicken > human > pig > cattle

and that, because of the dynamic nature of the partitioning, obtaining the best resolution between different cells requires joint consideration of the polymer composition and the time of sampling.

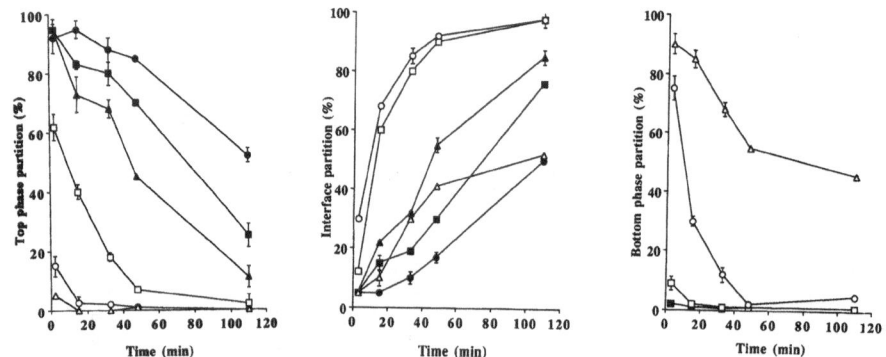

Figure 4. The time dependence of the top phase, interface and bottom phase partitioning of erythrocytes in 5D-4.17P low Δψ phase systems.(● rat, ■ turkey, ▲ chicken, □ human, ○ pig, △ cattle).

Microscopic examination of cells in the phase systems undergoing phase separation, or on samples removed at intervals and viewed on slides, showed that cells associated with the outer surface of dextran-rich phase droplets (see Figure 2 in Reference 6) and with the inner surface of PEG-rich phase droplets. As a consequence of this difference in orientation dextran-rich droplets were flocculated in the presence of cells, whereas PEG-rich droplets were not. The erythrocytes in Figure 3 and Figure 4 showed differences in their interaction with phase droplets. Rat, turkey or chicken erythrocytes sat proud on the dextran-rich phase droplets whereas pig and cattle cells attached firmly to the surfaces, changing cell shape, and following closely the curvature of the droplet surface. Human cells showed an intermediate behaviour.

Rat, turkey and chicken cells in the 5D-4.05P system were initially unattached to phase interfaces. The cells flowed freely within the PEG-rich streams in the developing bottom phase and were delivered into the developing top phase as these coalesced with the bulk interface. The pattern of phase separation appeared identical to that in cell-free systems with the slow coalescence mechanism predominating. However, later in partition time (about 65 min) chicken cells began to attach to the dispersion of microdroplets that was by then present, forming flocculated droplets by droplet-cell-droplet bridging, which sedimented faster than free droplets or free cells. As rat and turkey cells were still free at this stage a separation of their partition time course from that of the chicken cell arose. In the 5D-4.17P system the chicken erythrocytes began to attach between 5-15 min, the turkey cell at about 15-20 min and the rat cells between 30-40 min; times which correlate with the divergence of the cell partitions.

Human, pig and cattle cells were all attached to phase interfaces directly after mixing, and were not observed free in these systems. They increased the rate of coalescence (see Figure 5 for a schematic mechanism), readily seen by a decrease in $T_{immobile}$ compared with the cell free systems. Because of the increased rate of coalescence stream droplets were formed to which cells were attached. Cattle cells caused these stream droplets to coalesce to form larger stream droplets more readily than pig cells, whereas human cells had little effect. Once delivered to the top phase, these larger cattle cell-loaded droplets sedimented more rapidly than the pig and human cell-loaded droplets. Consequently the top phase partitions were in the order:

human > pig > cattle.

Human cells were differentiated from cattle and pig in these phase systems by clearly being less firmly attached to the phase interface than pig or cattle cells: they formed a diffuse band of cells lining the phase interfaces and sat prouder on dextran-rich phase droplets whereas cattle and pig cell formed tight bands of cell. (A relatively quantitative measure of this difference between pig and human erythrocytes has been obtained from contact angle measurements. [15,16] An additional feature of this weaker binding was that human cells were able to move on the surface of the phase interfaces; thus cell-capped droplets (large dextran-rich phase droplets in the top phase with cells accumulating as a cap on the upper end of the droplet as it sedimented), were seen for human cells but less frequently for pig and cattle cells.

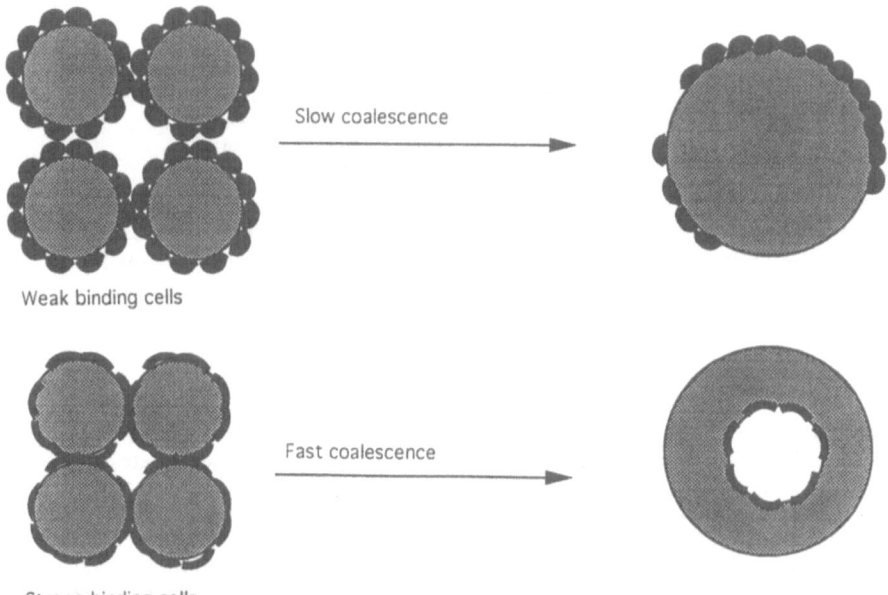

Figure 5. Cells influence the coalescence of phase droplets. Cells are shown on the surface of dextran-rich droplets. Depending on the strength of their attachment they promote slow or fast coalescence.

As the phase composition was moved further away from the critical point e.g., 5D-4.5P, all these cell were attached to the interfaces and the strength of binding increased.

It has been suggested that the molecular basis of differences in partitioning behaviour of erythrocytes of different species, in phase systems of low $\Delta\psi$, very close to the critical point, lies in differences in the lipid composition of their membranes. [6] We have found that cell surface carbohydrates are important determinants since treatment of erythrocytes with trypsin or with neuraminidase decreases the cell-phase interface interactions and increases top phase partition. Figure 6 shows this effect for cattle erythrocytes. Cell surface carbohydrate may contribute to determining cell-interface interactions in two ways: firstly, it is likely to be preferentially wetted by the dextran rich-phase rather than by the PEG-rich

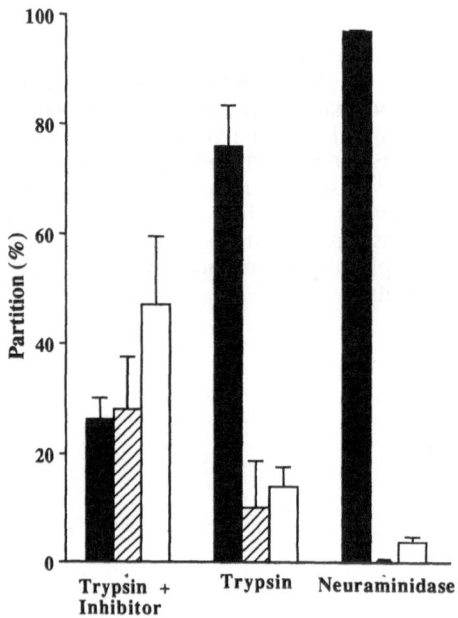

Figure 6. Effect of treatment of cattle erythrocytes with trypsin (in presence and absence of trypsin inhibitor) or neuraminidase on their partition in low $\Delta\psi$ phase system. Cells were partitioned in a 5D:3.5P system of for 8 min. Results are given for top phase, interface and bottom phase partitioning respectively.

phase and secondly, the carbohydrate chains might be expected to hinder access of the PEG-rich phase to wetting the outer lipid surface. Removal of carbohydrate chains might therefore be expected to increase the ability of the PEG-rich phase to interact with the cell surface. The removal of the sialic acid by neuraminidase was very effective in decreasing the cell-interface interaction. Since sialic acid is only one of a number of monosaccharides present in the glycocalyx, albeit the only charged (negatively) monosaccharide, it is difficult to explain this in terms of a decrease in hydrophilicity. Similarly, in phase systems of low $\Delta\psi$, it is difficult to explain this effect in terms of charge dependent interactions. One possible explanation might be that loss of these charge groups leads to alterations in the conformation of the carbohydrate chains, which permit increased interactions with the PEG-rich phase.

Affinity cell partitioning

Addition of PEG-palmitate (and other PEG-esters) to cells in phases of low $\Delta\psi$ phase systems increases their top phase partition.[17-18] It has been suggested that this depends on the extent of the hydrophobic interaction between the cell membrane and the palmitoyl residue.[19]

We find that PEG-palmitate decreases the cell-phase interface interaction, thereby increasing top phase partition. When erythrocytes of different species were partitioned in 5D-5P systems containing PEG-palmitate (Fig 7) time dependent partitions were observed that were comparable to those observed in a 5D-4.17P system in the absence of PEG-palmitate. The order of top phase partition was the same:

rat > turkey > chicken > human > pig > cattle

Rat, turkey and chicken cells were free from all interfaces early in the partition (1-3 min) whereas human, pig and cattle cells were attached, with human cells sitting prouder than

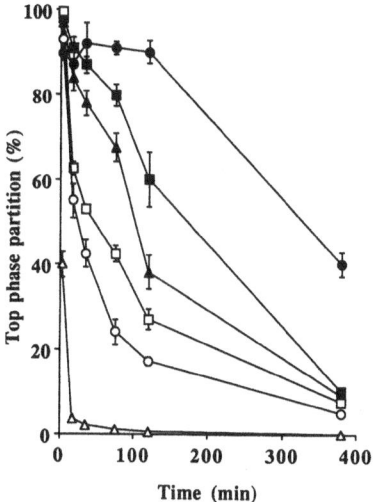

Figure 7. Influence of PEG-palmitate on the partitioning of erythrocytes in low $\Delta\psi$ phase system. Cells were partitioned in a 5D:5P phase system (2g) containing 20 ug of PEG-palmitate. (● rat, ■ turkey, ▲ chicken, ☐ human, ○ pig, △ cattle).

pig cells on the PEG-rich phase droplets (indicating a weaker association). Cattle cells differed from pig cells in that they showed a considerable bottom phase partition. Between 10-60 minutes all cells were attached to interfaces except rat cells, which finally attached to microdroplets after about 120 min.

In the absence of PEG-palmitate, none of the cells were present in the top phase because of the strong binding conditions present. Only by including PEG-palmitate to weaken the cell-interface interactions can top phase partitions be obtained, an effect equivalent to decreasing the interfacial tension. When the phase composition was increased, higher amounts of PEG-palmitate was required to produce similar partition behaviour (data not shown).

Charge-sensitive partitioning

The addition of certain salts which partition unevenly between the phases produce phase systems with an electrostatic potential difference between the phases. In the case of phosphate, the top phase is "positive relative to the bottom phase". Phase systems with polymer compositions some distance from the critical point can be used to provide top phase partitioning of cells, which have negative cell surfaces, which correlate with their charge status. They have thus been designated "charge-sensitive" phase systems. Typically 0.11M phosphate is used.

When we partitioned erythrocytes of different species is such systems we found that the mechanism underlying the increased top phase partition was due, as in the case of PEG-palmitate containing systems, to a decrease in the interaction of the cell with the polymer interfaces.

DISCUSSION

Phase separation depends critically upon the concentration of the polymers, their molecular weight and consequently upon the batches of polymers used. Nevertheless, the above descriptions of phase separation for the specific batches of PEG 6000 and Dextran T500 form the basis for a general description of phase separation.

The separation of two-phase systems of PEG and dextran into a PEG-rich upper phase and a dextran-rich bottom phase does not occur instantly but takes a significant time. Albertsson [1] has commented that "it depends not only on the difference in density between the phase and their viscosities, but also on the time needed for the small droplets formed during shaking, to coalesce into larger droplets". Our studies defining phase separation in term of the movement of complex microphases (droplets, stream, stream-droplets and multiple stream droplets) demonstrate that the coalescence of smaller droplets mentioned by Albertsson is complex.

Furthermore, phase separation is sensitively dependent on the concentration of the phase forming polymer: as polymer concentration is increased the speed of phase separation generally increases and the mechanism of phase separation shows a progressive change in the relative contribution of the microphases. This appears to arise from the interplay of two properties of the phase systems that depend critically on the polymer composition. Firstly, systems of increasing polymer composition fall on tie lines of increasing length i.e., they are of increasing distance from the critical point, and show an exponential increase in interfacial tension. [12] Since coalescence between droplets and with the bulk interface are processes that reduce the total interfacial surface area of the phase systems and decrease the surface energy, these processes will occur faster in phase systems of higher interfacial tension. Secondly, the density difference between the phases, and the difference in the two phases as solvents, increases with polymer composition. This feature contributes to the speed at which the microphases move relative to one another increasing with polymer composition. However, the increasing viscosity of the dextran-rich bottom phase with phase composition can impose a limit on the rate of phase separation, particularly by trapping PEG-rich streams and PEG-rich droplets in the bottom phase.

The mechanism by which phases separate can be summarised as follows. Shaking the phase system disperses the two phases into very small droplets; in the case of phases close to the critical point with the lowest interfacial tensions, these are so small that light is not scattered and the mixed systems appear clear. On standing the droplets begin to coalesce, increasing in size to give rise to turbidity and then the pitted appearance. The subsequent formation of streams depends on the phase composition (Figure 1). In phases close to the critical point the domains have little tendency to move and there is time for them to coalesce to form streams. Because coalescence is slow there is also time for the continuous phase between the coalescing droplets to drain away to form streams of the opposite phase, without becoming trapped with the coalescing domains. Therefore streams of both phase are formed, characteristically free of stream droplets. As the stream move slowly upwards and downwards the shear forces exerted on the streams are sufficient to overcome the relatively low interfacial tension and droplets are readily formed. Since the droplets have a low tendency to coalesce many are able to move to the bulk interface, with which they coalesce slowly. By contrast in phases further away from the critical point coalescence between adjacent domains occurs so rapidly that some of the continuous phase is trapped within the streams, forming stream droplets. Because the streams move rapidly upwards and downwards as phase separation quickly occurs, the droplets within the streams are flowing so fast in the same direction that they have reduced opportunities for coalescence. PEG-rich streams containing dextran-rich droplets on fusing with the bulk interface deliver the dextran-rich droplets into the top phase, where they eventually slow down and begin to sediment back to the bulk interface. Increased opportunities for coalescence now occur, aided by the high coalescence tendency of such systems, and larger droplets are formed. As these fall faster to the bulk interface, the mechanical drag forces cause them to adopt a pear shape and ultimately bud into two smaller droplets, which may subsequently coalesce with others. Consequently a competition develop between coalescence, which decreases the number of droplets and arises from the relatively high interfacial tension, and fragmentation/budding, which increases the number of droplets and which in this case arises from the rapid flow of the microphases. These process results in the turbidity clearing rapidly by the contraction of a discrete band of turbidity at the developing bulk interface, as streams and droplets rapidly arrive and coalesce with it, leaving increasingly clear phases above and below the contracting band. With increasing interfacial tension this banding mechanism becomes increasingly important and multiple phase droplets and rapid coalescence are increasingly seen.

Although aqueous two-phase systems have very low interfacial tensions compared with oil-water emulsion systems, [13] they show similarities in phase separation. For example, the clearance of dextran-rich droplets from the top phase and PEG-rich droplets from the bottom phase by upward and downward movements to the bulk interface respectively is similar to downward and upward creaming. [12] Furthermore, Becher has described multiple emulsions for oil-water systems (e.g., Figure 5-13 in reference [13]) rather like the multiple phase droplets we report here. Finally, the rapid coalescence mechanism suggested in Figure 1 to explain the formation of multiple phase droplets is similar to the process that Carrol describes for the inversion of oil in water emulsion, in which oil droplets coalesce so rapidly that droplets of the water continuum are trapped inside the oil droplets. [14]

Microscopic examination of the phase systems showed that the mechanistic basis for these differences in cell partitioning behaviour lies in differences in interaction between the cells and the phase interfaces that are formed in phase separation (which we describe above). As a consequence of this, as we shall describe, the cells influence the process of phase separation of the domains with which they are associated.

Cell partitioning arises as a consequence of the association of the cell with the surfaces of these complex microphases and can be summarised in terms of three types of binding conditions pertaining: weak, medium and strong (Figure 8).

In weak binding conditions cells are initially unattached to phase interfaces. They rapidly get into the top phase as free cells as phase separation continues, producing an initial high top phase partition. Subsequently they attach to dextran-rich phase droplets in the top phase and are delivered on these to the bulk interface. The bottom phase partition is low at all stages.

In medium binding conditions cells bind immediately to phase interfaces: in the top phase these are dextran-rich streams whereas in the bottom phase these are PEG-rich streams and dextran-rich stream droplets. In the top phase, cells attached to dextran-rich phase droplets appear from two sources: by fragmentation of dextran-rich streams and by delivery of dextran-rich stream droplets from the bottom phase. The top phase partition is initially high but declines as the cell-loaded phase droplets settle. The bottom phase partition is low at all times.

In strong binding conditions cells are attached to phase interfaces immediately and strongly accelerate phase separation. Cell loaded dextran-rich phase droplets in the top phase are rapidly cleared (seen as a contracting band in the partition tube), and the top phase partition falls rapidly. The bottom phase partition is high due to the delivery of cell-loaded PEG-rich stream droplets from the top phase and also due to the trapping of cell-loaded PEG-rich phase droplets in the bottom phase by the rapid delivery of viscous, dextran-rich phase from the top phase.

Cells can influence the rate of bulk phase separation, particularly if cell concentration is high and binding is strong. When cell concentration is relatively low, there is, naturally, little impact on bulk phase separation. However, it is important to appreciate that the partitioning of the cells is a consequence of their interactions with the process of phase separation.

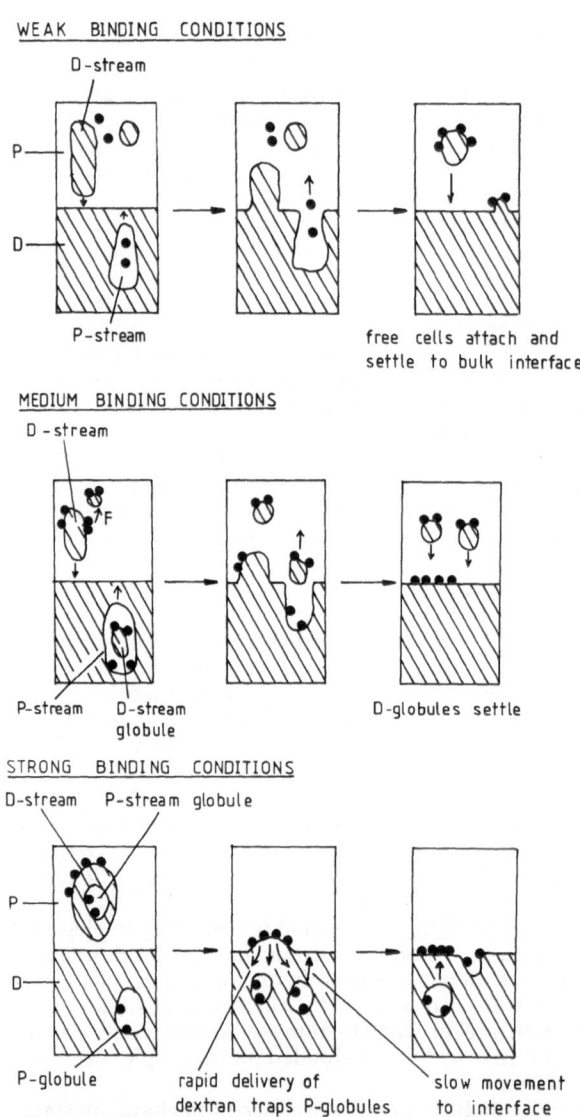

Figure 8. Behaviour of cells in phase systems of low $\Delta\psi$ under weak, medium and strong binding conditions. P designates PEG-rich phase domains and D designates dextran-rich phase domains. F indicates the fragmentation of a dextran-rich stream into a dextran-rich phase droplet.

Table 1 summarises the behaviour of the cells in the phase systems in terms of the type of binding observed. It demonstrate what has already been observed in Figure 3 and Figure 4, namely that a loss of resolution between different cell types will be observed as the polymer composition is increased or decreased from a critical concentration window. Taking the system too close to the critical point results in a decrease in cell-interface interaction to a degree that the cells do not influence the process of phase separation and so their partitions are not different. Moving the system too far from the critical point links cell partition to phase separation, but the high interfacial tension of the phase systems swamps the subtle differences in cell-interface interaction between the different cells and therefore, provides no discriminatory differences in partitioning behaviour.

Table 1. Interactions of erythrocytes of different species with microphases[1]

Erythrocytes	Phase system				
	5D:4.05P	5D:4.17P	5D:4.5P	5D:5P	5D:5.5P
Rat	weak	weak	weak	medium	strong
Turkey	weak	weak	medium	strong	strong
Chicken	weak	weak	strong	strong	strong
Human	weak	medium	strong	strong	strong
Pig	medium	strong	strong	very strong	very strong
Cattle	strong	strong	strong	very strong	very strong

[1]The behaviour of erythrocytes in low $\Delta\psi$ phase systems was classified as described in text and Fig 8.

This mechanism can be extended to affinity and charge sensitive phase partitioning. In the presence of PEG-ligands the cell surface becomes more PEG-like and top phase partition increases. Contact angle measurements have shown that the binding of PEG-palmitate by cells decreases the strength of their association with phase droplets, presumably because it increases their wettability by the PEG-rich phase. [15,16]

Thus the affinity partition has not arisen from the PEG-ligand partitioning preferentially into the top phase and "pulling the cell up"; in fact the partition coefficient of the PEG-ligand is probably irrelevant to the affinity partition as the cells are likely to be "coated" during mixing of the system. Although "species dependent" partitions are obtained (as in Figure 7) it does not necessarily follow that these arise from different affinities of the cells for PEG-palmitate. The order of top phase partitions obtained in the presence of PEG-palmitate is similar to that obtained in its absence in phases closer to the critical point (i.e., rat > turkey > chicken > human > pig > cattle). The principal effect of PEG-palmitate is to adsorb to the cell surfaces and weaken the cell-interface interaction so that the "separation window" is achieved. This view is supported by our observation that binding of PEG-palmitate to rat, chicken, human and pig erythrocytes appears very similar. A similar explanation may apply for the observations that PEG-iminoacetate-Cu(II) which binds primarily to histidine residues, produces a similar "species-ladder" in affinity partition. [20]

In systems that have a high $\Delta\psi$ the interaction with phase droplets is decreased on a charge-related basis. However, in both charge sensitive phase systems and PEG-palmitate/low $\Delta\psi$ systems it is important to appreciate that the partition obtained is the resultant of the charge status/magnitude of $\Delta\psi$ or binding of PEG-palmitate, and the affinity for the phase interfaces. Operationally for affinity and charge-dependent cell partition, phase composition is adjusted so that in the absence of the PEG-ligand or at low $\Delta\psi$ the top phase partition is very low. However, as we have shown, low partitions and normal partition times disguise a range of affinities for the interfaces.

Surface components may contribute to differing extents to these affinities. As we have seen, sialic acids make contributions both to the charge sensitive and non-charge sensitive partitioning.

It is clear from these studies that the interactions of cell with the phases systems that give rise to the partition occur early in the partitioning process, frequently when phase separation is just beginning.

REFERENCES

1. P.-Å. Albertsson. "Partition of Cell Particles and Macromolecules," Third Edition, John Wiley and Sons, New York (1986).
2. H. Walter, D.E. Brooks and D. Fisher, eds., "Partitioning in Aqueous Two-Phase Systems," Academic Press, Orlando, USA (1985).
3. D. Fisher and I.A. Sutherland.I.A., eds., "Separations Using Aqueous Phase Systems. Applications in Cell Biology and Biotechnology," Plenum Press, New York and London (1989).
4. D. Fisher, The separation of cells and organelles by partitioning in two-polymer aqueous phases, Biochem. J. 196:1 (1981).
5. H. Walter, E.J. Krob and D.E. Brooks, Membrane surface properties other than charge involved in cell separation by partition in polymer aqueous two-phase systems, Biochemistry 15:2959 (1976).
6. F.D. Raymond and D. Fisher, Partition of rat erythrocytes in aqueous polymer two-phase systems, Biochim. Biophys. Acta 596:445 (1980).
7. F.D. Raymond and D. Fisher, The effect of poly(ethylene glycol) palmitate on the partition of cells in aqueous polymer two-phase systems, Biochem. Soc. Trans. 8:118 (1980).
8. F.D. Raymond and D. Fisher, Cell surface properties and cell partition in two polymer aqueous phase systems. in:"Cell Electrophoresis in Cancer and other Clinical Research" A.W. Preece and P.A. Light, eds., Elsevier/North Holland Biomedical Press Amsterdam, New York, Oxford (1981).
9. D. Fisher and H. Walter, Cell separations and subfractionations by countercurrent distribution in two-polymer aqueous phase systems depends on non-equilibrium conditions, Biochim. Biophys. Acta 801:106 (1984).
10. D. Fisher, F.D. Raymond and H. Walter, Factors in cell separation by partitioning in two-polymer aqueous-phase systems, in:"Cell Separation Science and Technology," D.K. Kompala and P. Todd, eds., American Chemical Society, Washington DC (1991).
11. J. Ryden and P.-Å. Albertsson, Interfacial tension of dextran-polyethylene glycol-water two-phase systems, J. Colloid Interface Sci. 37:219 (1971).
12. H. Walter, F.D. Raymond and D. Fisher, Erythrocyte partitioning in dextran-poly(ethylene glycol) aqueous phase systems, J. Chromatogr. 609:219 (1992).
13. P. Becher, P. "Emulsions: theory and practice". Reinhold, New York (1965).
14. B.J. Carroll, The stability of emulsions and mechanisms of emulsion breakdown, in: "Surface and Colloid Science, vol 9," E. Matijwiuic, ed., Wiley Intercience, New York (1976).
15. B.N. Youens, D.W. Cooper, F.D. Raymond, P.S. Gascoine and D. Fisher, in: "Separations Using Aqueous Phase Systems. Applications in Cell Biology and Biotechnology," D.Fisher and I.A. Sutherland, eds., Plenum Press. New York and London (1989).
16. B.N. Youens, Cell partition and interfacial properties in two-polymer aqueous phase systems, Ph.D. Thesis. University of London (1986).
17. E. Eriksson, P.Å. Albertsson and G. Johansson, Hydrophobic surface properties of erythrocytes studied by affinity partition in aqueous two-phase systems, Molecular and Cellular Biochem. 10:123 (1976)
18. H. Walter, E.J. Krob and R. Tung, Hydrophobic affinity partition in aqueous two-phase systems of erythrocytes from different species. Systems containing polyethylene glycol-palmitate, Exp. Cell Res. 102:14 (1976).
19. H .Walter, Surface properties of cells reflected by partitioning: red blood cells as a model, in: "Partitioning in Aqueous Two-Phase Systems", H. Walter, D.E. Brooks and D. Fisher, eds., Academic Press, Orlando (1985).
20. H. Walter, K.E. Widen and G. Birkenmeier, Immobilized metal ion affinity partitioning of erythrocytes from different species in dextran-poly(ethylene glycol) aqueous phase systems, J. Chromatogr. 641:279 (1993).

PROTEIN REFOLDING USING CHAOTROPIC AQUEOUS TWO-PHASE SYSTEMS

Timothy A. Spears and Daniel Forciniti

Chemical Engineering Department
University of Missouri-Rolla
Rolla, MO 65401

INTRODUCTION

In this report we explore novel aqueous two-phase systems and their applications to the recovery of recombinant proteins, which are expressed as insoluble and inactive protein aggregates (inclusion bodies).

Protein aggregation is effectively a side product of protein folding[1] since as a protein folds toward its native structure it can also go through alternative folding pathways forming intermediates that aggregate. As in the folding process, the aggregation process is strongly dependent on temperature, pH, salt type, and salt concentration.[2-7] In addition, the aggregation process is strongly dependent on protein concentration with high protein concentrations favoring the formation of aggregates.[5] Protein folding is catalyzed in some bacteria by enzymes (chaperones) which bind intermediates along the folding pathway thereby avoiding protein association. Similarly, hydrophobic probes and poly(ethylene)glycol, PEG, reduce the formation of protein aggregates during *in vitro* protein refolding by the same mechanism.

The formation of inclusion bodies provides a convenient and clean alternative for protein purification, **if** the active protein can be recovered from the protein aggregates. Under ordinary circumstances, the inclusion bodies formed by the eukaryotic protein contain almost no impurities. Thus, by separating the inclusion bodies from the cell debris by centrifugation[8] or ultrafiltration,[9] it is possible to obtain a highly pure protein in a single purification step. After the inclusion bodies have been isolated, they are suspended in a buffer and exposed to denaturing conditions to solubilize the protein. The denaturing agent is then diluted (either by dialysis or by adding an appropriate buffer) to refold the protein. Unfortunately, during the refolding process the proteins re-aggregate. Thus, only a fraction of the active protein is recovered.[10]

Several routes have been explored to prevent the re-aggregation of proteins and thus improve the recovery of the active protein. It has been found that by refolding protein at very low concentrations or in the presence of large amounts of denaturant, the recovery of native protein increases. These procedures have two disadvantages: (1) the refolding of the protein

Aqueous Biphasic Separations: Biomolecules to Metal Ions
Edited by R.D. Rogers and M.A. Eiteman, Plenum Press, New York, 1995

occurs at a very low rate, and (2) the refolded protein is obtained at very low concentrations. Inclusion of a hydrophobic probe, e.g., a polypeptide, in the refolding mixture to enhance protein recovery has also been investigated. The hydrophobic probe reduces the number of aggregates by interacting with some intermediates (e.g., molten globules) in the folding reaction. Unfortunately, hydrophobic probes are specific for each particular protein; a more general technique is highly desirable. Other chemicals normally used to improve refolding are EDTA, oxidized glutathione, glutathione, sodium octanoate, and sodium palmitate. Some of these compounds such as glutathione, oxidized glutathione, or disulfide isomerase, are added to obtain the correct pairs of cysteines.[11] Cleland and Wang[12] have found that by adding a cosolvent like PEG to the refolding mixture, the formation of aggregates is minimized and the recovery of the active protein is enhanced. Cleland and Wang have successfully used PEG (3 to 30 grams/liter) for the recovery of carbonic anhydrase II. This approach is especially appealing because it seems independent of the kind of protein. However, it produces a highly dilute protein solution contaminated by large amounts of denaturant. Clearly, improvements to this technique are necessary before it can become attractive commercially.

We are developing a novel protein refolding technique in which aqueous two-phase PEG/salt systems are used to both dissociate the inclusion bodies and refold the protein. The general idea behind this technique is the following. It is well known that when aqueous solutions of PEG and salt are mixed together they phase separate,[13] and that proteins dissolved in the resulting two-phase systems partition unequally between the phases due to solubility differences.[13] In our refolding technique, proteins are dissolved in an aqueous PEG/salt solution which has been chosen to have the following characteristics: (1) proteins are conformationally stable in the PEG phase (where PEG acts to prevent aggregation), (2) protein aggregates are dissociated in the salt phase (where the salt acts as a denaturant), and (3) proteins tend to partition into the PEG phase (driven by differences in solubility).

We have recently reported some preliminary work which confirms the feasibility of the proposed refolding scheme.[14] We found that a series of chaotropic salts forms two liquid phases with PEG when spiced with NaCl. We demonstrated that systems containing PEG and a mixture of NaSCN and NaCl were capable of simultaneously dissociating aggregates and refolding carbonic anhydrase II. Between 20 and 30% of the total enzymatic activity was recovered in the PEG-rich phase, indicating that the technique should be investigated further.

In this paper we present further experimental evidence showing that aqueous two-phase systems can be used for the refolding of recombinant proteins. First, we briefly describe our findings using PEG/NaCl/NaSCN systems to refold carbonic anhydrase II. Second, we describe the properties of a new phase system whose phase forming species are guanidine hydrochloride (GuHCl), sodium chloride, and PEG. Finally, we present some preliminary results for the refolding of carbonic anhydrase II in these novel phase systems.

EXPERIMENTAL

Materials

PEG-8000 and carbonic anhydrase II (Lot # 30H9490), ribonuclease A (Lot # 11H8142), and pig hemoglobin (Lot # 118F9315) were obtained from Sigma (St. Louis, MO, USA). All the other chemicals were of analytical-reagent grade.

Methods

Preparation of the Phase Systems. Four and ten gram phase systems were prepared by mixing PEG with bi-distilled water and salts in the appropriate amount and stirring for several minutes.[15] The pH was adjusted by adding either phosphate or tris buffers at a concentration of 67 mmol and 50 mmol respectively. Duplicates were prepared for each phase system. Most of the systems considered in this work are in the one phase region at room temperature. Therefore, the systems are driven into the two-phase region by increasing the temperature.

Phase diagrams were determined as follows. The phase systems prepared as indicated above were placed into a stirred cell of a temperature controlled spectrometer and the absorbance at 450 nm was measured as a function of temperature. First, we increased the temperature until a sharp increase in the absorbance was observed indicating that the system had become cloudy. The temperature at which the systems became cloudy was recorded as the cloud point temperature (forward experiments). Second, we decreased the temperature until the absorbance observed at room temperature was recovered. The temperature at which the system became transparent was recorded as the cloud point temperature (backward experiments). The cloud point temperature obtained by the backward and forward experiments agrees within 1 °C.

Partition Experiments. For the partition experiments, 0.25, 0.5, or 1 mL of a 2.5 mg/mL protein stock solution was added to the phase systems replacing an equal amount of buffer. The addition of the protein to the phase system was done following different protocols which are described in the results section. The systems were stirred for several minutes and placed into a water bath to induce the phase transition. After complete phase separation (between 1 to 3 hours), samples from the top and bottom phases were taken carefully with a micropipet. After dilution, the absorbance at 280 nm was measured using quartz cuvettes in a double beam Hitachi spectrophotometer (model U-2000). The calculated extinction coefficient for the native protein was 1.90 mg/mL, while the extinction coefficient for the denatured protein was 1.65 mg/mL. The enzymatic activity was measured by recording the increase of absorbance at 348 nm due to the hydrolysis of p-nitrophenylacetate.[10] The activity of the enzyme was determined as the slope of the absorbance at 348 nm versus time plot. Since p-nitrophenylacetate hydrolyzes even in the absence of the protein, a blank was discounted to account for the changes in the reactant. The enzymatic activity of the fresh enzyme was approximately 2 units/mg of protein at 20 °C.

Protein Denaturation. Carbonic anhydrase II was denatured by incubating the enzyme for 24 h in a 5 M solution of guanidine hydrochloride either at room temperature or at 37 °C.

PEG Concentration. PEG concentration was determined using a colorimetric assay. 5 mL of 0.5 M perchloric acid was added to 1 mL of PEG solution. After 15 minutes the precipitate was discarded. 1.0 mL of 5% BaCl and 0.4 mL of 0.1 M iodine were added to 4.0 mL of the solution. After 15 minutes the absorbance was measured at 535 nm. A calibration curve was made using PEG solutions of known concentration.

NaSCN Concentration. NaSCN concentration was determined using a colorimetric assay. 4.0 mL of 1000 ppm iron reference solution was diluted to 40 ppm with water. 3.0 mL of this solution was added to 1.0 mL of NaSCN solution and the absorbance was measured at 480 nm. A calibration curve was made using NaSCN solutions of known concentrations. No interference by PEG or NaCl was detected in the range of concentrations studied.

Chloride Concentration. Cl⁻ and SCN⁻ were titrated using mercuric nitrate with diphenylcarbazone as the indicator. The amount of Cl⁻ was calculated by subtracting the amount of SCN⁻ determined by the method described above.

RESULTS AND DISCUSSION

PEG/NaCl/NaSCN Systems

In a previous article,[14] we have reported that NaSCN spiced with NaCl forms two liquid phases with PEG at reasonable temperatures. In addition, we found that the mutual solubility of PEG/NaCl/NaSCN systems exhibits a minimum at 25 °C.[14]

We determined phase diagrams, and we studied the effect that the total concentration of PEG and the pH have on the cloud point temperature for PEG/NaCl/NaSCN systems. We found that the amount of total salt in each phase is rather high. A high concentration of salt in the bottom phase is desirable because it will favor the dissolution of the aggregates. On the contrary, a high concentration of salt in the top phase is undesirable because it will inhibit the refolding of the protein.

Partition experiments were performed with native carbonic anhydrase II and with carbonic anhydrase II aggregates. The protein partition coefficient, $K = C^{top}/C^{bottom}$, was always very high. The protein mass balance, however, did not close. This indicates some adsorption of the protein at the interface. The interface of the system containing a high concentration of NaSCN becomes turbid and very broad when 1 mL of the stock protein solution is added but remains clear when 0.5 mL of the stock protein solution is added. This indicates that the composition of the phases depends on the amount of protein present in the system. Our findings suggest that the usual assumption that the phase systems are not altered by the addition of low protein concentrations is not valid for these systems.

We recovered enzymatic activity in the top phase of the systems containing relatively small amounts of NaSCN.[14] We used the PEG/NaCl/NaSCN systems to dissolve carbonic anhydrase aggregates and to refold the protein. First, we suspended the aggregates in different amounts of tris buffer. The aggregates did not dissolve and no enzymatic activity was observed. The aggregates were then placed into the selected phase systems. The aggregates dissolved almost instantaneously when the systems were still in the one phase region and no enzymatic activity was detected. The systems were placed in a water bath at 40 °C to induce the phase separation. Samples from both the top and bottom phases were analyzed for protein content and enzymatic activity. In the system containing 30% NaSCN no activity was recovered in the bottom phase but 0.386 units/mg were recovered in the top phase. We did not find any protein in the bottom phase, indicating that the protein accumulates at the interface.

As indicated above, the protein precipitated at the interface in PEG/NaCl/NaSCN systems. Therefore, only a fraction of the total activity was recovered in the top phase while no activity was recovered in the bottom phase. Since carbonic anhydrase II is denatured by NaSCN solutions as low as 5% w/w, we argued that the amount of NaSCN in the top phase was too high to allow the protein to refold. In this work, we present experimental evidence which confirms our previous explanations. In Table 1 we report the composition and volumes of the top and bottom phases of a series of PEG/NaCl/NaSCN systems. The following conclusions can be drawn:

(1) The amount of PEG in the top phase is rather high (from 35.5% w/w in system A to 53.2% w/w in system D). In this range of PEG concentrations proteins may precipitate and accumulate at the interface. The amount of PEG needed to obtain two phases may be decreased by manipulating the pH and temperature. Changes in pH and temperature may also be beneficial by increasing the solubility of the protein in the PEG-rich phase.

(2) Our results indicate that the amount of NaSCN in the PEG-rich phase exceeds the amount expected if both NaCl and NaSCN partition in the same proportion. Also, the concentration of NaSCN in the PEG-rich phase is high enough to completely denature the enzyme. However, some enzymatic activity is recovered in the PEG-rich phase of PEG/NaCl/NaSCN systems. We believe that part of the NaSCN is actually bound to the PEG

backbone. As a consequence of this binding, NaSCN is sequestered from the bulk, allowing the protein to refold.

(3) The data can be, within experimental error, plotted in a PEG/total salt or PEG/NaCl plane to represent phase diagrams. We refrain to present the data in this form because: (a) it is not possible to move along the experimental tie lines to change the volume ratio between top and bottom phases. Moreover, the volume of the PEG-rich phase increases with increasing amounts of PEG; which is contrary to well established observations;[15] (b) the resulting "binodal" does not separate the plane into one and two phase regions; i.e., we can only use these curves when the concentrations are close to the determined cloud points.

(4) The amount of PEG and of NaSCN in the PEG-rich phase increases with increasing temperature. On the contrary, the amount of NaCl in the PEG rich phase decreases with increasing temperature.

Table 1. Volumes and phase compositions of several PEG/NaCl/NaSCN systems.[a]

System	Total Composition (w/w)			Volume (mL)		PEG (w/w)		NaSCN (w/w)		NaCl (w/w)	
	PEG	NaSCN	NaCl	Top	Bottom	Top	Bottom	Top	Bottom	Top	Bottom
A	20	4.5	10.5	5.5	3.2	35.5	0.9	6.3	0.75	7	18
B	22	5.1	11.9	4.8	4	45.4	0.5	7.7	0.9	5.5	22.9
C	20	4.5	10.5	4.3	4.4	46.1	0.9	7.2	1.2	4.8	19.5
D	22	5.1	11.9	4.1	4.4	53.2	0.4	7.3	1.2	3.4	23.7
E	20	4.6	10.5	4.8	4.3	41.6	0.5	5.7	1.13	3.6	19.1
F	22	5.1	11.9	4.3	4.5	50.9	0.2	6	1.2	--	--
G	22	5.1	11.9	5.3	3.5	40.5	1.4	7.7	2.3	7.6	21.7
H	24	5.8	13.3	4.5	4.2	52.8	0.5	10.6	2.1	5.2	25.5
I	22	5.2	11.9	4.7	4.2	45.7	0.7	8.8	2.6	6.4	21.6
J	24	5.7	13.3	4.5	4.3	51	3	10.1	2.2	3.5	25.9

[a]Systems A, B, C, D, E, and F contain phosphate buffer, systems G, H, I, and J contain tris buffer. Systems A, B, G, and H are at room temperature, C and D are at 46 °C, E and F are at 67 °C, and I and J are at 41 °C.

PEG/NaCl/Guanidine Hydrochloride Systems

After proving that some chaotropic salts such as NaSCN form two liquid phases with PEG and that the resulting aqueous two-phase systems can be used for the simultaneous dissolution of protein aggregates and refolding of the protein, we examined a more conventional chaotropic salt used for protein refolding: guanidine hydrochloride, as a phase forming species with PEG.

Figure 1 shows the cloud point coordinates at room temperature and pH = 9.4 (phosphate buffer 67 mmol) for PEG/NaCl/GuHCl systems. For a given combination of GuHCl and PEG concentrations, the figure gives the concentration of NaCl needed for obtaining two phases. Some general trends can be drawn from this figure. First, the higher the PEG concentration, the smaller the amount of NaCl needed to obtain two phases for any GuHCl concentration. Second, at low GuHCl concentrations, the amount of NaCl needed to obtain two phases depends on the amount of PEG in the system. Finally, at very high PEG concentrations, the cloud point location appears independent of the NaCl concentration.

Since a protein can unfold, refold, or form aggregates, depending on the concentration of guanidine hydrochloride, we investigated the phase behavior of PEG/NaCl/GuHCl systems as a function of the ratio of guanidine hydrochloride to sodium chloride in the system.

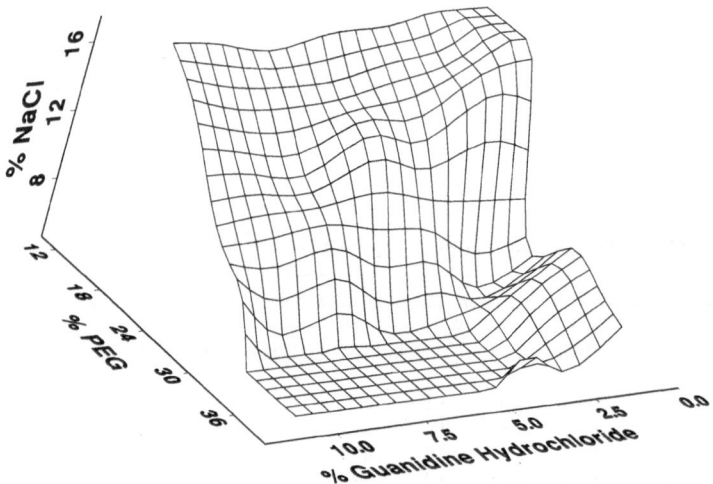

Figure 1. Cloud points of PEG/NaCl/GuHCl systems at room temperature.

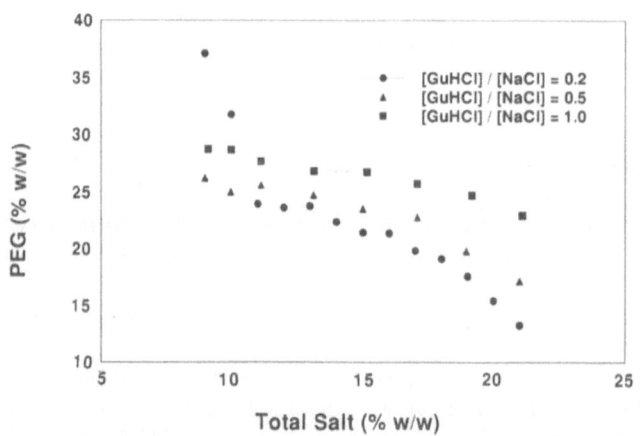

Figure 2. Concentration of PEG at the cloud point versus total salt concentration for different [GuHCl]/[NaCl] ratios (T = 25 °C, pH = 9.4).

Figure 2 shows a plot of the concentration of PEG versus the total salt concentration at the cloud point for different [GuHCl]/[NaCl] ratios. The figure shows that, except at low total salt concentrations, the concentration of PEG needed to obtain two liquid phases increases with increasing GuHCl to NaCl ratio.

We also investigated the effect that the pH has on the cloud point temperature. Figure 3 shows the cloud point temperature versus the ratio of GuHCl/NaCl in the system at different pHs at total PEG and salt concentrations of 20% w/w and 15% w/w respectively. The figure shows that for phosphate buffers the cloud point temperature decreases with increasing pHs. The figure also shows that there is almost no difference between the two systems containing tris buffer at two different pHs. Therefore, our data indicate that the cloud point temperature depends not only on the pH, but also on the type of buffer used.

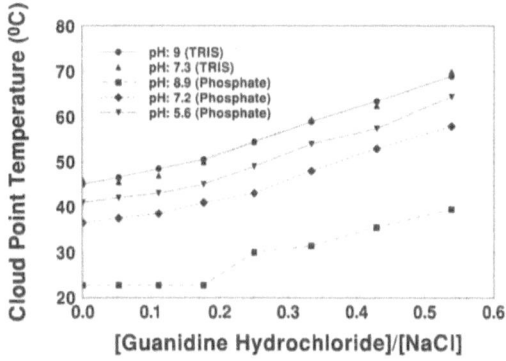

Figure 3. Cloud point temperature versus the ratio of [GuHCl]/[NaCl] at different pHs (total PEG concentration = 15% w/w; total salt concentration = 20% w/w).

One of our goals is to concentrate the refolded protein in a small volume to simplify the subsequent purification steps. Therefore, we needed to identify phase systems whose top phases represent a small fraction of the total volume. In the search for these optimum volume ratios we must: (1) keep the phase separation temperature within reasonable limits; (2) keep the total amount of PEG low to allow the proteins to move into the top phase and without being precipitated by PEG; and (3) keep the total amount of GuHCl within practical limits. The boundaries for these three restrictions depend on each particular protein. The results of our search for the optimum volume ratio are summarized in Figure 4. Figure 4 shows the volume fraction of the bottom phase (V_{bottom}/V_{total}) versus the ratio of GuHCl to NaCl in the system at different pHs. The figure shows that the volume of the bottom phase decreases with increasing amounts of GuHCl. Therefore, high total GuHCl concentrations in the systems increase the volume of the top phase, which is undesirable. The figure also shows that the volume fraction of the bottom phase increases with increasing pH for a given GuHCl/NaCl ratio. Therefore, basic pHs should be used.

We can summarize our findings as follows: (1) the cloud point temperature decreases with increasing pHs, total salt concentrations, and PEG concentrations; (2) the volume of the top phase increases with increasing PEG concentrations and with increasing ratios of GuHCl to NaCl; and (3) the volume of the top phase decreases with increasing pH.

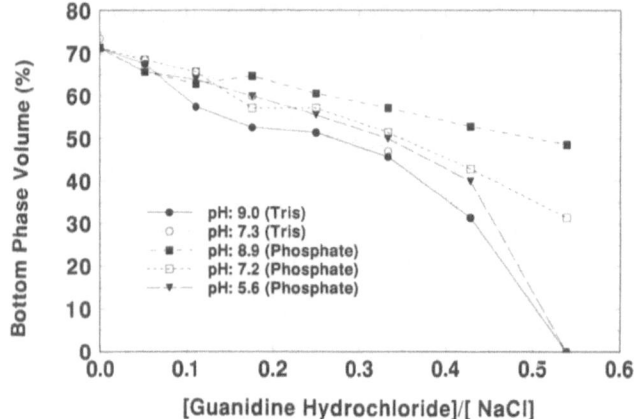

Figure 4. Bottom phase volume fraction versus the ratio of GuHCl to NaCl at different pHs.

Partition Experiments

One difficulty encountered with PEG/NaCl/NaSCN systems was the relative instability of the phases. In PEG/NaCl/NaSCN, changes in volume ratios, and phase separation temperature were observed when small amounts of proteins were added to these systems. Therefore, before testing the PEG/NaCl/GuHCl systems with carbonic anhydrase II, we performed some partition experiments using hemoglobin and ribonuclease to monitor the response of these systems upon the addition of proteins. Table 2 shows the partition coefficients of bovine ribonuclease and pig hemoglobin at six different pHs in a system containing 10% w/w PEG, 4.3% w/w GuHCl, and 20.8% w/w NaCl. Both proteins are strongly rejected by the bottom (salt-rich) phase.

Table 2. Partition coefficients of ribonuclease and hemoglobin in PEG/NaCl/GuHCl systems at different pHs.

pH	Partition Coefficient	
	Ribonuclease	Hemoglobin
5.5[a]	2.33	7.47
7.6[a]	1.51	1.62
8.8[a]	1.70	3.01
9.4[a]	1.72	4.11
7.4[b]	1.37	2.25
8.9[b]	1.35	2.78

[a] Phosphate buffers. [b] Tris buffers.

The rejection of the proteins by the bottom phase is a desirable property for the design of the refolding protocol. In addition, we notice that the partition coefficients of both ribonuclease and pig hemoglobin in phosphate buffer are higher than in tris buffer of a similar pH. PEG/NaCl/GuHCl systems appear quite stable upon addition of moderate amounts of proteins; i.e., no observable changes in cloud point temperature or in phase compositions were detected. This observation opposes our observation with PEG/NaCl/NaSCN systems.

Refolding Experiments

After the PEG/NaCl/GuHCl systems were well characterized, we studied the partitioning and refolding of carbonic anhydrase II in these systems. Table 3 lists the partition coefficient, the volume ratio, and the activity of carbonic anhydrase II in the top and bottom phases and in the refolding step (see below) for each of the systems studied. This set of experiments can be grouped into three categories: (1) systems containing tris buffer which phase separate at 51 °C (systems A and B); (2) systems containing phosphate buffer which phase separate at 39 °C, in which the refolding of the protein is done in the presence of large amounts of PEG (systems C and D); and (3) systems containing phosphate buffer which phase separate at 39 °C, in which the refolding of the protein is done in the presence of small amounts of PEG (systems E to J).

Table 3. Activity of carbonic anhydrase in PEG/NaCl/GuHCl systems.

System	K	V_T/V_B	Activity		
			Refolding Step	Top Phase	Bottom Phase
A	98	0.3	N/A	0.0	0.0
B	14.6	0.3	0.6	0.0	0.0
C	2.0	--	0.6	0.5	3.0
D	9.5	--	0.9	0.17	3.1
E	11.0	0.7	--	0.4	2.5
F	15.2	0.6	2.4	0.05	0.0
G	11.3	0.6	2.2	0.13	0.0
H	100	0.8	0.76	0.0	0.0
I	15.0	0.9	2.6	0.23	0.0
J	14	0.7	2.2	0.23	0.0

The composition of systems A and B is 17% w/w PEG, 17% w/w NaCl, 3% w/w GuHCl, and 2.5 mg of the denatured protein. The pH was adjusted using tris buffer at pH = 7.5. The difference between systems A and B is the method of contacting the components of the phase system. In system A, all of the materials are added at time zero and a sample is taken to check enzymatic activity. The system is then placed into the water bath to induce phase separation. In system B, the denatured protein is contacted with a buffer containing the total amount of PEG (refolding step). After one hour, a sample is taken to check enzymatic activity. This activity is reported in column four of Table 3. NaCl is added and the test tube is placed into a water bath to induce phase separation. There was no activity recovery in either phase of any of the systems and only a small recovery in the refolding step. Two possible reasons are: (1) the temperature needed to induce phase separation is too high to allow refolding of the protein, and (2) the amount of PEG in the refolding step is too high.

To decrease the phase separation temperature we used a series of systems containing sodium phosphate buffer which split into two-phases at 39 °C (systems C and D). The composition of systems C and D is 17% w/w PEG, 16% w/w NaCl, 3% GuHCl, and 2.5 mg of protein. System C contains native protein and system D contains denatured protein. The pH was adjusted using phosphate buffer at pH = 9.9. The native protein has a moderate partition coefficient while the denatured protein partitions strongly to the top phase where some enzymatic activity is recovered. The very high specific enzymatic activity in the bottom phase suggests that the denatured protein partitions into the top phase while the native protein partitions into the bottom phase. Although the concentration of GuHCl in the refolding step is not high enough to denature the protein, the native protein looses most of its activity during this step. Moderate concentrations of GuHCl combined with high concentrations of PEG seem to partially inactivate the enzyme.

Following Cleland and Wang,[12] we mixed the denatured protein with a small amount of PEG (approximately 3 g/L) and allowed the protein to refold. After one hour, phase separation was induced by adding the other components and increasing the temperature. We followed this protocol using three different phosphate buffers (systems E, F, I, and J, pH = 9.38; system G, pH = 7.6; system H, pH = 5.5) and using three different protein concentrations (system I contains 2.5 mg of protein, system J contains 1.25 mg of protein, and system K contains 0.75 mg of protein). All these systems contain 10% w/w PEG and 20.8% NaCl. System E contains the native enzyme while systems F to J contain the denatured enzyme. Systems F to H contain 4.2% w/w GuHCl, system I contains 2.1% w/w, and system J contains 1.05% w/w. The concentration of GuHCl in the refolding mixture was 0.64 M in systems F to H, 0.32 M in system I, and 0.17 M in system J. Following this contacting protocol, we were able to recover full activity during the refolding step except at pH = 5.5 (system H). The recovery of activity during the refolding step is nearly independent of protein concentration in the range studied by us (compare systems F, I, and J). However, the enzymatic activity observed in the top phase after phase separation was very low in all cases. Also, the partition coefficients of the refolded protein are always larger than those of the native protein. These results were unexpected since the result of the refolding step should be the native protein, which should partition accordingly. However, the native protein and the protein that has been previously denatured and refolded in the presence of PEG exhibit different partition behavior, indicating that some changes have occurred during the refolding process. According to Cleland and Wang,[12] PEG binds to the first intermediate of the folding reaction of carbonic anhydrase II. They argue that PEG is released from the protein surface after the protein is folded. Our findings are in partial disagreement with these arguments. We can explain the differences in partition behavior between the native and refolded protein arguing that PEG does not completely detach from the protein surface after the protein is folded and therefore, acts as an affinity ligand driving the protein into the top phase. Once the protein is in the top phase, the amount of PEG is so large that the protein precipitates. Unfortunately, we were not able to analyze the precipitate which appears shortly after the phases separate.

CONCLUSIONS

Aqueous two-phase systems have been widely used for the purification of proteins. One widely recognized advantage is that aqueous two-phase systems offer a benign environment for most proteins. In this work, we have developed aqueous two-phase systems containing salts which denature proteins with the aim of using them for the recovery of proteins from inclusion bodies.

We have presented detailed information about the phase behavior of systems containing PEG, NaCl, and either NaSCN or guanidine hydrochloride. We have studied the effect that the ratio of chaotropic to non-chaotropic salt, pH, temperature, and amount of PEG has on these phase systems.

We have used these two-phase systems to study the refolding of carbonic anhydrase II. We have found that under some conditions it is possible to recover a fraction of the total activity in the PEG-rich phase. Unfortunately, the amount of PEG in the PEG-rich phase is so high that the protein precipitates. We believe that these phase systems can be used to further our understanding of cosolvent assisted protein refolding. We are presently testing our protocol with hen lysozyme and with inclusion bodies of the same protein.

RERERENCES

1. T.E. Creighton, Protein folding, *Biochem. J.* 270:1 (1990).
2. A. Mitraki and J. King, Protein folding intermediates and inclusion body formation, *Bio/Technology* 7:690 (1989).
3. A. Light, Protein solubility, protein modifications and protein folding, *BioTechniques* 3:302 (1985).
4. B. Fisher, I. Summer, and P. Goodenough, Isolation, renaturation, and formation of disulfide bonds of eukaryotic proteins expressed in *Escherichia coli* as inclusion bodies, *Biotech. and Bioeng.* 41:3 (1993).
5. J.M. Donovan, G.B. Benedek, and M.C. Carey, Self-association of human apolipoproteins A-I and A-II and interactions of apolipoprotein A-I with bile salts: quasi-elastic light scattering studies, *Biochemistry* 26:8116 (1987).
6. P.J. Tichy, F. Kapralek, and P. Jecmen, Improved procedure for a high yield recovery of enzymatically active recombinant calf chymosin from *Escherichia coli* inclusion bodies, *Prot. Expr. Pur.* 4:59 (1993).
7. D.N. Brems, Solubility of different folding conformers of bovine growth hormone, *Biochemistry* 27:4541 (1988).
8. F.A.O. Marston, The purification of eukaryotic polypeptides synthesized in *Escherichia coli, Biochem. J.* 240:1 (1986).
9. S.M. Forman, E.R. DeBernardez, R.S. Feldberg, and R.W. Swartz, Crossflow filtration for the separation of inclusion bodies from soluble proteins in recombinant *Escherichia coli* cell lysate, *J. Membrane Sci.* 48:263 (1990).
10. A. Ikai, L. Tanaka, and H. Noda, Reactivation kinetics of guanidine denatured carbonic Anhydrase B, *Arch. Biochem. Biophys.* 190:39 (1978).
11. G. Georgiou and G.A. Bowden, Inclusion body formation and the recovery of aggregated recombinant proteins, *in*: "Recombinant DNA Technology and Applications," Chapter 12, McGraw-Hill, New York (1991).
12. J.L. Cleland, S.E. Builder, J.R. Swartz, M. Winkler, J.Y. Chang, and D.I.C. Wang, Polyethylene glycol enhanced protein refolding, *Bio/Technology* 10:1013 (1992).
13. P.-Å. Albertsson. "Partitioning of Macromolecules," J. Wiley & Sons, New York (1986).
14. D. Forciniti, Protein refolding using aqueous two-phase systems, *J. Chromatogr. A*, 668:95 (1994).
15. D. Forciniti, C.K. Hall, and M.-R. Kula, Influence of polymer molecular weight and temperature on phase composition of aqueous two-phase systems, *Fluid Phase Equilibria* 61:243 (1991).

PARTITIONING OF POLYETHYLENE GLYCOL(PEG)-PROTEIN CONJUGATES IN PEG/DEXTRAN AQUEOUS TWO-PHASE SYSTEMS

Cristina Delgado

Molecular Cell Pathology
Royal Free Hospital School of Medicine
Rowland Hill Street
London NW3 2PF
England, U.K.

SUMMARY

Covalent attachment of polyethylene glycol (PEG) to proteins produces an increase in their partition coefficient (K) in PEG/dextran aqueous two-phase systems. At least for bovine serum albumin (BSA) and granulocyte-macrophage colony stimulating factor (GM-CSF) a linear relationship exists between log K of the conjugate and the number of PEG molecules attached to it (up to 20 PEG molecules for BSA and 3 for GM-CSF). The proportionality constant (i.e., the increment in log K per PEG molecule attached) is protein specific. For PEG-GM-CSF conjugates, the proportionality constant increases with the concentration of the polymers (PEG and dextran up to 6%, w/w) in the two-phase system. A further increase in the polymer concentration up to 6.5% leads to a decrease in the proportionality constant. To explain these observations it is proposed that the interaction between the PEG-modified conjugate and the polymers in the phase system includes a positive interaction between the PEG attached to the PEG-modified conjugate and the PEG coils in the top phase, in addition to the excluded volume interactions generally accepted as the main interactions determining the partitioning of native proteins.

INTRODUCTION

The partition of proteins in aqueous two-phase systems of PEG and dextran is known to be influenced by protein surface properties such as charge, hydrophobicity-hydrophilicity as well as by the protein molecular size.[1-3] Chemical modification of the protein surface conveys changes to the protein's partitioning behavior and this has been exploited to produce ligands for affinity partitioning.[4,5] Attachment of PEG to proteins or triazine dyes increases their partitioning to the top PEG-rich phase[6,7] whereas attachment of dextran increases their partitioning to the bottom dextran-rich phase.[8]

Aqueous Biphasic Separations: Biomolecules to Metal Ions
Edited by R.D. Rogers and M.A. Eiteman, Plenum Press, New York, 1995

167

The molecular mechanisms leading to such partitioning behavior are, however, poorly understood. Several models for the interaction of the protein with the phase forming polymers are being constructed, exploiting different thermodynamic approaches to predict the partitioning of native proteins.[9,10] A common feature that emerges from several of them is that the interaction of the protein with the PEG present in the top phase of PEG/dextran two-phase systems includes excluded volume interactions in addition to a weak attraction between the polymer coils and the protein.[11] Progress has been somewhat slower on the mechanisms leading to the changes in partitioning upon chemical modification of proteins. This is in part due to the lack of experimental data to address the influence of the degree of modification on the partitioning behavior. In this paper we report the partitioning behavior of PEG-BSA and PEG-GM-CSF conjugates as a function of the degree of PEG-modification. The influence of the polymer concentration in the phase system on the partitioning of PEG-GM-CSF conjugates of increasing degree of modification is also studied. A model for the interactions between the PEG-modified conjugates and the polymers in the phase system is proposed and its implications on the potential application of phase partitioning for the analysis of PEG-protein conjugates are discussed.

MATERIAL AND METHODS

Materials

Chemicals were obtained from the following sources: PEG (Mr 6000, BDH, Poole, U.K.), Dextran T-500 (Pharmacia, Sweden), MPEG (Mr 5000, Union Carbide, USA), tresyl chloride (Fluka, Switzerland), fluorescamine (Sigma, Poole, U.K.). All the other reagents were Analar grade from BDH (Poole, U.K.). BSA was from Sigma and ^{125}I-GM-CSF from Amersham (contamination with free iodine was negligible and no release of free iodine was observed during storage at 4°C for the recommended period after which preparations were discarded). A kit for protein determination using the Coomassie brilliant blue assay was from Pierce (Rockford, Illinois, USA).

Coupling of PEG to proteins

Monomethoxypolyethylene glycol (MPEG) of 5000 Mr was activated with tresyl chloride essentially as previously described.[12] MPEG is selected as the parent polymer to provide only one active end in each molecule and thus avoids the formation of cross-linked products. The activated polymer was then incubated with the relevant protein in coupling buffer at room temperature for 2 h using a rotary mixer and subsequently stored at 4°C until processed. Coupling buffer was PBS from Gibco, except for BSA where 0.05 M sodium phosphate pH 7.5 containing 0.125 M sodium chloride was substituted. Although the polymer attached to the proteins is MPEG, the conjugates are referred to as PEG-modified proteins for simplicity.

Phase partitioning

For bovine serum albumin, a two-phase system consisting of 4.75% PEG-6000, 4.75% dextran T-500, 0.15 M sodium chloride and 0.01 M sodium phosphate buffer pH 6.8 (non-charge sensitive system) was prepared on a weight for weight basis (1 g total) from stock solutions of 40% (w/w) PEG, 20% (w/w) dextran, 0.44 M sodium phosphate buffer pH 6.8, 0.6 M sodium chloride, distilled water, and 0.1 g of solutions of either the native BSA or the

PEG-modified BSA. After 30-40 inversions, the mixture was left to settle at room temperature until complete separation of the phases was observed (15 to 20 min). Top and bottom phases were then analyzed for protein concentration by the Coomassie Brilliant blue assay. The partition coefficient is defined as the ratio between the protein concentrations in the top and the bottom phases.

For ^{125}I-GM-CSF, two-phase systems of PEG-6000, dextran T-500 (see concentrations at the figure legend), 0.15 M NaCl and 0.01 M sodium phosphate were prepared in advance from the stock solutions of 20% (w/w) dextran T-500, 40% (w/w) PEG-6000, 0.5 M sodium phosphate pH 6.8, and 0.6 M sodium chloride. After mixing, the system was allowed to settle into the top PEG-rich phase and the bottom dextran-rich phase at room temperature. The top and bottom phases were stored in separated bottles at 4°C until required. Before use, the stored phases were allowed to equilibrate to room temperature and the biphasic system reconstituted by mixing top and bottom phases at a 1:1 volume ratio. Aliquots of reaction mixtures (1 µL) or FPLC fractions (50 µL) were incorporated and the system mixed by 30 to 40 gentle inversions. After separation of the phases, aliquots from top and bottom phases were analyzed for ^{125}I levels. The partition coefficient (K) is calculated as the ratio between ^{125}I levels in top and bottom phases.

Estimation of free amino groups

A fluorescamine (4-phenylspiro[furan-2(3H),1'-phthalan]-3,3'-dione) assay, as described by Stocks et al.,[13] was used. Fluorescamine interacts with the primary amino groups of proteins (lysine residues and amino terminus) to form a fluorophore (390 nm excitation, 475 nm emission). The resulting fluorescence is proportional to the amine concentration and is not affected by the presence of TMPEG in the sample.[13]

Gel permeation chromatography

Samples (200 µL) were chromatographed on a Pharmacia FPLC system fitted with a Superose-12 HR 10/30 column previously equilibrated with PBS. The samples were loaded on to the column and then eluted at a flow rate of 0.3 mL per minute; 0.25 mL fractions were collected. The elution buffer was sterile PBS. Protein concentration in the fractions was established by the levels of ^{125}I quantified in a gamma counter. Molecular weight markers at concentrations of 3 mg/mL were run similarly and the fractions analyzed for protein content using a Coomassie Brilliant blue assay from Pierce. Molecular weight markers include beta-amylase (200,000 Da), alcohol dehydrogenase (yeast, 150,000 Da), bovine serum albumin (66,000 Da), carbonic anhydrase (29,000 Da), cytochrome c (12,400 Da), and aprotinin (6,500 Da).

RESULTS

BSA was incubated with increasing concentrations of TMPEG, the PEGylating agent, for 2 h at room temperature in a rotary mixer. The covalent attachment of the polymer to the protein was monitored by partitioning in PEG/dextran aqueous two-phase systems and by the number of amino groups available for reaction with fluorescamine (i.e., free amino groups). Figure 1 (left panel) shows that there is reduction in free amino groups which conveys increased partition coefficient in the PEG/dextran aqueous two-phase selected. Thus exposure of BSA to TMPEG leads to attachment of the polymer to amino groups and the resulting conjugates have increased partition coefficient in the two-phase system of PEG and dextran. The disappearance of available amino groups is a quantitative measure of the PEG molecules attached (N_{PEG}) according to the following relationship:

$$N_{PEG} = f * (NH_{2(total)} - NH_{2(free)}) \qquad (1)$$

where f represents the proportional contribution of amino groups to the total nucleophiles in the protein reacting with TMPEG. (This relationship holds until complete reaction of one of the nucleophile types if more than one is involved.) The relationship between the number of PEG molecules attached (calculated from (1) assuming that amino groups are the only nucleophiles involved, i.e., $f = 1$) and the partition coefficient (K) in the PEG/dextran two-phase system is shown in Figure 1 (right panel). There is a linear relationship between log K and the number of PEG molecules attached (Figure 1, right panel). The slope of the curve (0.079 ± 0.01, $\pm$ SE) provides an estimate for the increment in log K per PEG molecule attached. Note that if other nucleophiles are involved in the reaction with TMPEG then the calculated increment in log K per PEG molecule is in fact an overestimate.

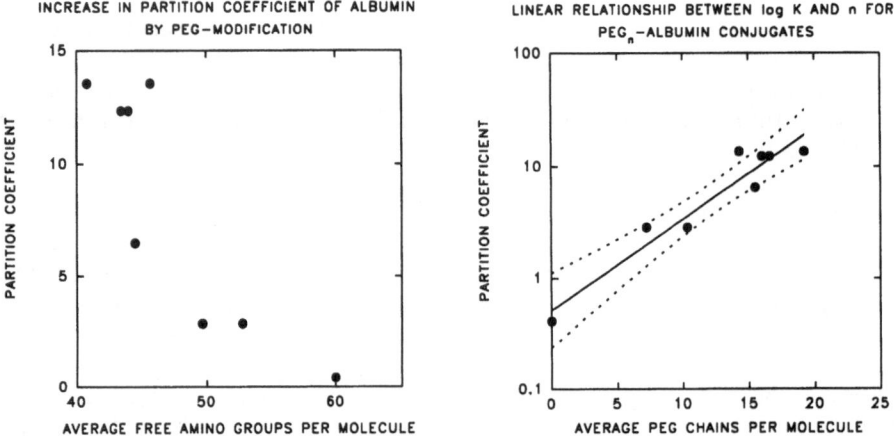

Figure 1. Left panel: Partition coefficient of PEG-modified bovine serum albumin as a function of the average free amino groups per molecule (biphasic system: 4.75% PEG-6000, 4.75% dextran T-500, 0.15 M sodium chloride, 0.01 M sodium phosphate pH 6.8). Right panel: Linear relationship between log K and average degree of modification for PEG-BSA conjugates. The solid line represents the linear regression (slope = 0.079 ± 0.01, intercept = -0.28 ± 0.16, estimate ± SE) and the dashed lines the 95% confidence interval.

PEG-[^{125}I]GM-CSF conjugates containing 1 to 3 PEG molecules were produced by incubation of [^{125}I]GM-CSF with TMPEG (molar excess of TMPEG to NH_2 of 37.5:1 and 75:1) followed by fractionation by gel permeation chromatography in a Superose 12 column.[14] There was an increase in partition coefficient in a PEG/dextran two-phase system with the degree of modification (Figure 2, left panel). Again a linear relationship is found between log K and the number of PEG molecules attached (Figure 2, right panel).

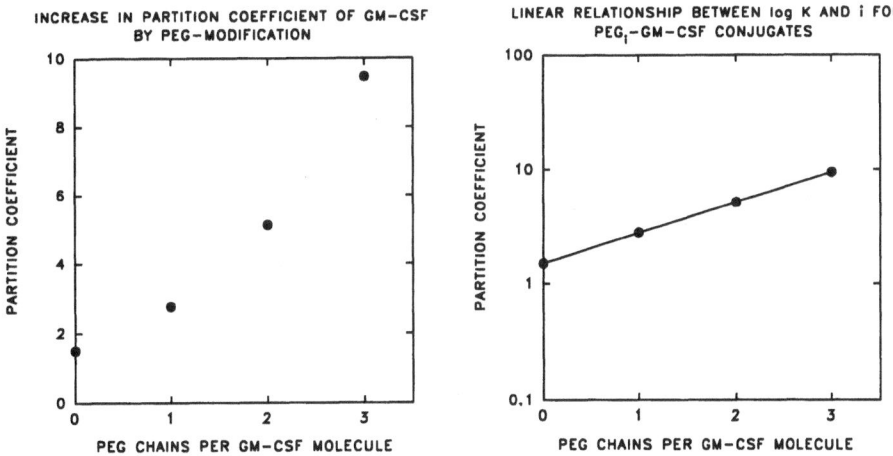

Figure 2. Left panel: Partition coefficient of PEG-[^{125}I]GM-CSF as a function of the number of PEG chains per molecule (biphasic system: 6% PEG-6000, 6% dextran T-500, 0.15 M sodium chloride, 0.01 M sodium phosphate pH 6.8). Right panel: Linear relationship between log K and degree of modification for PEG-[^{125}I]GM-CSF conjugates. The line shows the linear regression (slope = 0.267 ± 0.008, intercept = 0.175 ± 0.017, estimate ± SE).

The influence of the polymer concentration of the phases on the partitioning of PEG-modified conjugates was investigated with PEG-GM-CSF. Preparations of increasing average degree of modification were obtained by incubation of [^{125}I]GM-CSF with increasing concentrations of TMPEG. The partition coefficient of the resulting reaction mixtures was analyzed in PEG/dextran aqueous two-phase systems with polymer concentrations ranging from 4.75% (w/w) to 6.5% (w/w) (Figure 3). In all biphasic systems tested, the [^{125}I]-labelled material showed increased partition coefficient (and thus increased average degree of modification) with the molar excess of TMPEG used in the coupling reaction (Figure 3, all panels). The polymer concentration of the phases had little influence on the partition coefficient of unmodified [^{125}I]GM-CSF (Figure 3, top-left panel) and [^{125}I]GM-CSF incubated with relatively low molar excess of TMPEG, c.f. 10 (Figure 3, middle-left panel) and 37.5 (Figure 3, bottom-left panel). The PEG-[^{125}I]GM-CSF conjugates in the reaction mixtures prepared at a molar excess of TMPEG of 75 showed increased partition coefficient with the concentration of polymers of the phases over the whole range studied (Figure 3, top-right panel). The PEG-[^{125}I]GM-CSF conjugates in the reaction mixtures prepared at molar excess of TMPEG of 150 (Figure 3, middle-right panel) and 305 (Figure 3, bottom-right panel) showed increased partition coefficient with the concentration of polymers of the phases up to 6% and then a reduction in the partition coefficient with a further increase in the polymer concentration of the phases up to 6.5%. (Note that only 1 μL of reaction mixture was added to the phase system and therefore the contribution of TMPEG to the total polymer concentration of the system is negligible.) Thus the polymer concentration of the phases influences only the partition coefficient of PEG-GM-CSF and the direction of the change in partition coefficient (increase or reduction) depends upon the polymer concentration of the phases and the degree of PEG-modification of the conjugates.

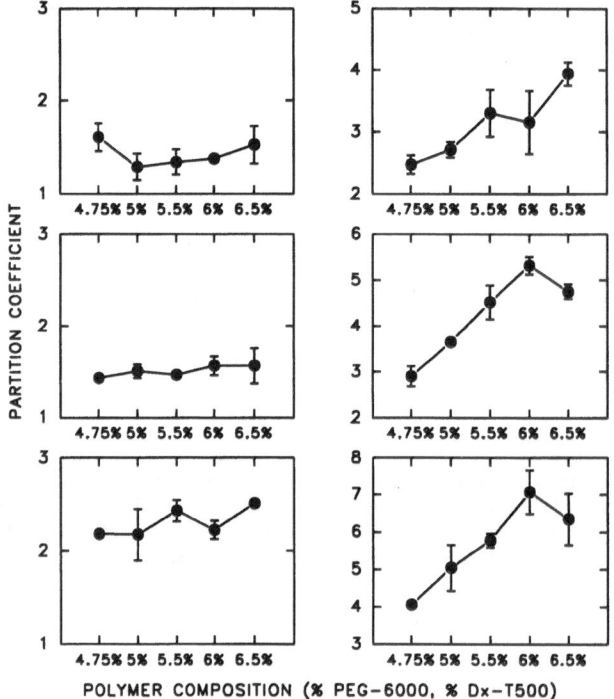

Figure 3. Influence of the two-phase system composition (polymer concentration) on the partition coefficient of ^{125}I-GM-CSF in reaction mixtures prepared by incubation of ^{125}I-GM-CSF and TMPEG at TMPEG/NH$_2$ ratios of 10 (middle-left panel), 37.5 (bottom-left panel), 75 (top-right panel), 150 (middle-right panel) and 305 (bottom-right panel). The top-left panel shows the result for unmodified [^{125}I]GM-CSF.

The relationship between log K and number of PEG molecules attached was then studied as a function of the polymer concentration of the phases (Figure 4). The average degree of modification in reaction mixtures prepared at molar excess of TMPEG of 10, 37.5, 75, 150 and 305 was calculated from the relationship between log K and number of PEG molecules attached per GM-CSF molecule known for the biphasic system 6% PEG/6% dextran (Figure 2, right panel and figure legend). In all two-phase systems tested a linear relationship between log K and the average number of PEG molecules attached to PEG-GM-CSF conjugates was found (Figure 4). As anticipated from the results seen in Figure 3, the slope of the curves increased when the polymer concentration was raised up to 6% (Figure 4 and figure legend). A further increase in the polymer concentration up to 6.5% produced a decrease in the slope of the curve (Figure 4 and figure legend).

In order to get insight into the mechanisms leading to the partitioning behavior observed, a plot of the increment in log K per PEG molecule (slope of the curves in Figure 4) versus the concentration of PEG in the top phase of the biphasic system was constructed (Figure 5). This showed that the increment in log K is not proportional to the concentration of PEG in the top phase (Figure 5).

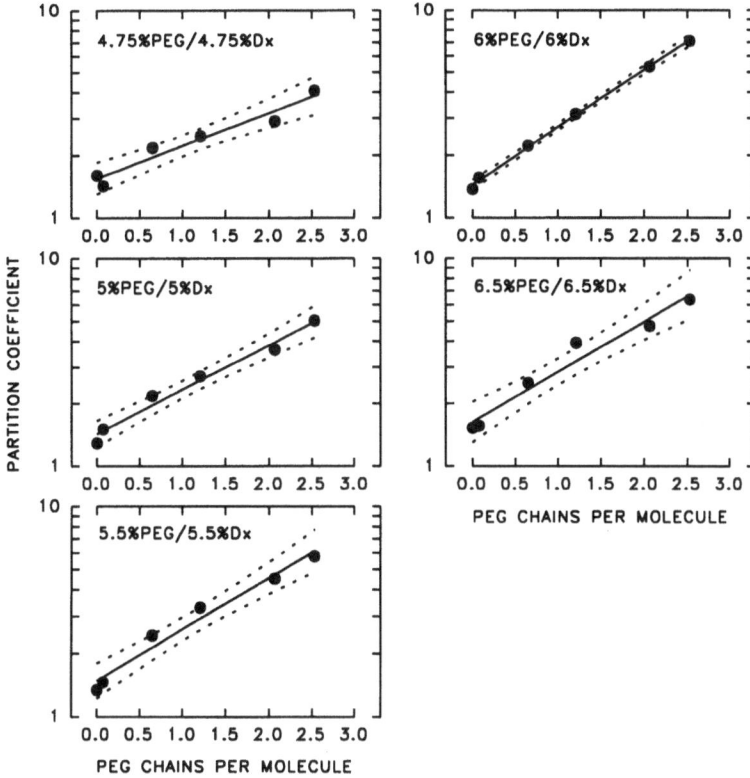

Figure 4. Linear relationship between log K and average PEG chains per GM-CSF molecule in two-phase systems of the following concentrations (%, w/w) of PEG-6000 and dextran T-500: 4.75% (top-left panel), 5% (middle-left panel), 5.5% (bottom-left panel), 6% (top-right panel), 6.5% (middle-right panel). The solid lines show the following linear regressions (estimate ± SE): system 4.75%, slope = 0.155 ± 0.019, intercept = 0.192 ± 0.044; system 5%, slope = 0.214 ± 0.016, intercept = 0.154 ± 0.037; system 5.5%, slope = 0.243 ± 0.021, intercept = 0.171 ± 0.048; system 6%, slope = 0.267 ± 0.008, intercept = 0.175 ± 0.017; system 6.5%, slope = 0.242 ± 0.024, intercept = 0.212 ± 0.056.

DISCUSSION

The covalent attachment of PEG to BSA and GM-CSF conveys an increase in their partition coefficient, K, in PEG/dextran aqueous two-phase systems. The addition of each PEG molecule (at least up to 20 for BSA and 3 for GM-CSF) produces a constant increment in log K. The increment in log K per PEG molecule added is smaller for PEG-BSA conjugates (0.079 ± 0.01, see legend to Figure 1) than for PEG-GM-CSF conjugates (0.155 ± 0.019, see legend to Figure 4) in a biphasic system containing 4.75% PEG and 4.75% dextran, i.e., increment in log K is protein specific. The different approaches used to establish the relationship between log K and PEG molecules attached cannot account for the protein-specific behavior. Note that if other nucleophiles in BSA had participated in the reaction, i.e., $f < 1$, then the increment in log K per PEG molecule would have been smaller. The different experimental approaches used to include the protein into the biphasic system (see material and methods) certainly had an influence in the volume ratio between top and bottom phases. However, this should not influence the polymer composition of the resulting top and bottom phases and, therefore, the partition coefficient of the macromolecules.

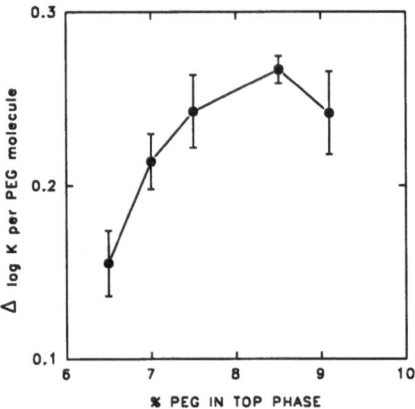

Figure 5. Increment in log K per PEG molecule as a function of the concentration of PEG in the top phase of the biphasic system. The latter were obtained from the phase diagram of the dextran D48-polyethylene glycol 6000 published by Albertsson.[2]

  In an attempt to explain these observations we have considered a simple model for the partitioning mechanism in which two types of interactions between the PEG-modified conjugate and the polymers in the phases are involved. The first interaction is a positive interaction between the PEG molecules attached to the conjugate and the PEG coils in the top phase. The second interaction between the PEG-modified conjugate and the polymers in the phases is of the type of excluded volume effects. The positive interaction between the PEG attached to the conjugate and the PEG in the top phase is the driver for the increase in partition coefficient with the number of PEG molecules attached (without this type of interaction it is difficult to explain the increased partition coefficient). It seems reasonable to presume that this type of interaction should be dependent on the amount of PEG attached and be rather protein independent. The second type of interaction, dominated by excluded volume effects (negative interaction), opposes the positive interaction between the attached PEG and the PEG in the top phase. Since the partition of macromolecules in PEG/dextran aqueous two-phase systems is mainly (but not exclusively) dominated by excluded volume effects[9,11] it is reasonable to postulate that this type of interaction should play a significant role in the partitioning of PEG-modified conjugates. The increase in size upon attachment of each PEG molecule (relative to the size of the native protein) determines the magnitude of this second interaction and therefore influences the change in log K. This simple model seems to be sufficient to explain the protein specific increment in log K per PEG molecule attached reported in this paper. This will occur if the addition of PEG molecules to two different proteins has different proportional effects on the changes in log K imposed by each of the two types of interactions involved in the partitioning: the positive interaction that depends on the mass of PEG attached to the protein and the negative interaction that depends on the increase in size of the conjugate.

The linearity between log K and the number of PEG molecules is easily anticipated if both interactions produce changes in log K which are linear with respect to the number of PEG molecules attached. It is well established that for globular proteins there is a linear relationship between log K and molecular size (with negative slope).[15] Therefore, if the addition of each PEG molecule produces a constant increment to the size of the conjugate, there will be a constant relative decrease in log K per PEG molecule added, opposing the increment due to the PEG induced positive interaction. This implies that a linear increase in log K per PEG molecule occurs as a result of the positive interaction if an overall linear relationship between log K and the number of PEG molecules is maintained. Notably, for PEG-GM-CSF conjugates the addition of 1 to 3 PEG molecules produced similar increments in hydrodynamic radius per PEG molecule (Selisko et al., manuscript in preparation). This suggests that at least for PEG-GM-CSF conjugates with 1 to 3 substitutions, the positive interaction between the attached PEG and the PEG in the top phase produces a change in log K which is linearly related to the amount of PEG attached. Whether this is a general feature for all PEG-modified conjugates or specific for PEG-GM-CSF remains to be established. One could, for example, envisage situations where the first PEG molecule attached and subsequent PEGs have a different impact on size. In the example where PEG-BSA conjugates were analyzed, mixtures containing conjugates of various degrees of modification were produced. SDS-PAGE and gel permeation chromatography failed to resolve the mixtures in individual conjugates and therefore no information on the increment in molecular size per PEG molecule is available. There is a possibility that the log K for PEG-modified conjugates increases linearly with the number of PEG molecules attached only when the resulting conjugate increases its size by the same value with each PEG molecule.

The increase in increment in log K with the concentration of polymers in the phases (see Figure 5) indicates that the positive interaction between the attached PEG and the PEG in the top phase produces an increase in the log K of the conjugate, since the increase in concentration of polymers in the phases produces a decrease in log K as a result of increased excluded volume effects.[9] The lack of proportionality between the increment in log K and the concentration of PEG in the top phase of the system (see Figure 5) could result if either only one of the interactions or the two interactions give rise to non-proportional changes in increment in log K per PEG molecule attached. Unfortunately, there is no information available on the changes in log K with the concentration of polymers in the system for proteins of different sizes and therefore further investigation is required to address this point.

If the log K of a PEG-modified conjugate is proportionally related to the number of PEG molecules attached only when the size of the conjugate increases by a certain value, then phase partitioning could be exploited not only as an analytical tool to establish the degree of modification,[14] but also to make predictions on the likely increase in molecular size upon addition of each PEG molecule. The latter is especially important, since it has bearings in the *in vivo* biodistribution and pharmacokinetics of PEG-protein conjugates. For many of the PEG-protein conjugates produced to date (see reference 16) the increment in size per PEG molecule could not be measured with other classical techniques such as SDS-PAGE and gel permeation chromatography.

ACKNOWLEDGMENTS

I would like to thank Dr. G.E. Francis and Dr. D. Fisher for their comments and support. Financial support from the Cancer Research Campaign (U.K.) is acknowledged.

REFERENCES

1. H. Walter, D.E. Brooks, and D. Fisher (eds.). "Partitioning in Aqueous Two-Phase Systems, Theory, Methods, Uses and Applications to Biotechnology," Academic Press, New York (1985).
2. P.A. Albertsson. "Partition of Cell Particles and Macromolecules," John Wiley and Sons, New York(1986).
3. D. Fisher and I.A. Sutherland (eds.). "Separations Using Aqueous Phase Systems. Applications in Cell Biology and Biotechnology," Plenum Press, New York and London (1989).
4. A. Gordes, J. Flossdorf, and M.R. Kula, Affinity partitioning: development of a mathematical model describing behavior of biomolecules in aqueous two-phase systems, *Biotechnol. Bioengineer.* 30:514 (1987).
5. J.M. Harris, E.C. Struck, M.G. Case, M.S. Paley, M. Yalpani, J.M. van Alstine, and D.E. Brooks, Synthesis and characterization of polyethylene glycol derivatives, *J. Polym. Sci.* 22:341 (1984).
6. K.A. Sharp, M. Yalpani, S.J. Howard, and D.E. Brooks, Synthesis and application of a poly(ethylene glycol)-antibody affinity ligand for cell separations in aqueous polymer two-phase systems, *Anal. Biochem.* 154:110 (1986).
7. G. Kopperschlager and G. Johansson, Affinity partitioning with polymer-bound Cibacron blue F3G-A for rapid, large-scale purification of phosphofructokinase from Baker's yeast, *Anal. Biochem.* 124:117 (1982).
8. G. Johansson and M. Joelsson, Affinity partitioning of enzymes using dextran-bound procion yellow HE-3G. Influence of dye-ligand density, *J. Chromatogr.* 393:195 (1987).
9. N.L. Abbott, D. Blankschtein, and T.A. Hatton, On protein partitioning in two-phase aqueous polymer systems, *Bioseparation* 1:191 (1990).
10. D. Forciniti and C.K. Hall, Theoretical treatment of aqueous two-phase extraction by using virial expansions. A preliminary report, *in*: "Downstream Processing and Bioseparations: Recovery and Purification of Biological Products," ACS Symposium Series, Vol. 419, J.-F.P. Hamel, J.B. Hunter, and S.K. Sikdar, eds., American Chemical Society, Washington D.C. (1990).
11. N.L. Abbott, D. Blankschtein, and T.A. Hatton, Protein partitioning in two-phase aqueous polymer systems. 1. Novel physical pictures and a scaling-thermodynamic formulation, *Macromolecules* 24:4334 (1991).
12. C. Delgado, J.N. Patel, G.E. Francis, and D. Fisher, Coupling of poly(ethylene glycol) to albumin under very mild conditions by activation with tresyl chloride: characterization of the conjugate by partitioning in aqueous two-phase systems, *Biotechnol. Appl. Biochem.* 12:119 (1990).
13. S.J. Stocks, A.J. Jones, C.W. Ramey, and D.E. Brooks, A fluorometric assay of the degree of modification of protein primary amines with polyethylene glycol, *Anal. Biochem.* 154:232 (1986).
14. C. Delgado, F. Malik, B. Selisko, D. Fisher, and G.E. Francis, Quantitative analysis of polyethylene glycol (PEG) in PEG-modified proteins/cytokines by aqueous two-phase systems, *J. Biochem. Biophys. Methods*, in press (1994).
15. S. Sasakawa and H. Walter, Partition behavior of native proteins in aqueous dextran-poly(ethylene glycol)-phase systems, *Biochemistry* 11:2760 (1972).
16. C. Delgado, G.E. Francis, and D. Fisher, Uses and properties of PEG-linked proteins, *in*: "Critical Reviews in Therapeutic Drug Carrier Systems," CRC Press, Boca Raton, Florida, (1992).

USE OF THE AQUEOUS TWO-PHASE PARTITION TECHNIQUE FOR CHARACTERIZATION AND QUALITY CONTROL OF RECOMBINANT PROTEINS

Boris Y. Zaslavsky

Ameritest and Research Company, Inc.
5475 Perkins Rd.
Bedford Heights, OH 44146

Partitioning in aqueous polymer two-phase systems is widely recognized as a highly efficient, versatile, and cost-effective method for separation and purification of biological materials.[1-6] Aqueous two-phase systems arise in aqueous mixtures of different water-soluble polymers or a single polymer and a specific salt. When two certain polymers, e.g., dextran and polyethylene glycol are mixed in water above certain concentrations, the mixture separates into two immiscible aqueous phases. There is a clear interfacial boundary and one phase is rich in one polymer and the other phase is rich in the other. The aqueous solvent in both phases provides media suitable for biological products.

When a solute is put into such a system, it distributes between the two phases. The procedure is very simple and similar to extraction. Solutions of two polymers are mixed and a two-phase system is formed. To speed phase settling, centrifugation may be used. Partition behavior of a solute may be influenced by many variables, such as pH, polymer and salt composition of the system, temperature, and so on. Partitioning of a solute is characterized by the partition coefficient K defined as the ratio between the concentrations of the solute in the two phases.

Numerous applications of the technique to bioseparation are based on the idea that the partition behavior of a given solute is solute-specific and independent of the presence of other solutes in the aqueous two-phase system employed (provided there are no intermolecular interactions between the solutes). Thus, the partition coefficient K for a given solute in the two-phase system of a fixed composition is constant and specific for the solute.

It has also been shown in numerous studies that the partition coefficient K is highly sensitive to minor changes in the solute structure. For example, elimination or replacement of a single residue in a protein's structure may lead to dramatic changes in the K value. An illustrative example is offered by the results reported by Raymond et al.[7] Partition coefficient values for two membrane-bound alkaline phosphatase isoforms differing by a single phosphate group were significantly different in aqueous Dextran-poly(ethylene glycol) two-phase systems, while the electrophoresis on gradient gels, could not distinguish between the isoforms.

Aqueous Biphasic Separations: Biomolecules to Metal Ions
Edited by R.D. Rogers and M.A. Eiteman, Plenum Press, New York, 1995

Partitioning of β-lactoglobulins A and B in the aqueous poly(ethylene glycol)-3400/potassium phosphate two-phase system was reported[8] to be different. These two proteins differ by two amino acid residues: β-lactoglobulin A has aspartic acid and valine, while β-lactoglobulin B has glycine and alanine at residue positions 64 and 118, respectively. The proteins have an isoelectric point difference of only 0.1 pH unit. The partition coefficients for β-lactoglobulins A and B in the aforementioned system amount to 0.08 and 0.04, respectively.[8] Partition coefficients of horse and pig insulins differing by a single amino acid residue (at position 9) in the same system were also found[8] to be different, 21.2 and 19.4, respectively. Chemical modifications of proteins by direct chemical treatment[9] or genetic engineering techniques[10,11] were shown to change the protein partitioning significantly. It should also be mentioned that protein conformational changes may affect its partition behavior dramatically.[12,13]

Numerous similarities between fundamentals of solute partitioning in aqueous polymer two-phase systems and in common solvent systems, such as the octanol-water system, have been found.[6] It has been established by Zaslavsky *et al.*,[14-16] in particular, that the logarithm of the partition coefficient of a solute in aqueous Dextran-poly(ethylene glycol) and Dextran-Ficoll two-phase systems is a measure of the solute relative hydrophobicity. The hydrophobicity of a solute is defined[17] as a measure of the intensity of molecular interactions of the solute with water in dispersed systems, the dispersing medium in which is water. The generally accepted measure of the solute hydrophobicity is the logarithm of the solute partition coefficient in the octanol-water two-phase system, log P.[18,19] This characteristic (log P) is widely used as the general descriptor of the solute structure in quantitative structure-activity relationships (QSAR) analysis for conventional drugs. The logarithm of the solute partition coefficient in an aqueous two-phase system, ln K, represents the relative hydrophobicity of the solute more adequately than the log P value for several reasons described in detail in reference 6. The one obvious advantage of ln K as the relative hydrophobicity measure over the log P parameter is that it may be applied to biological macromolecules.

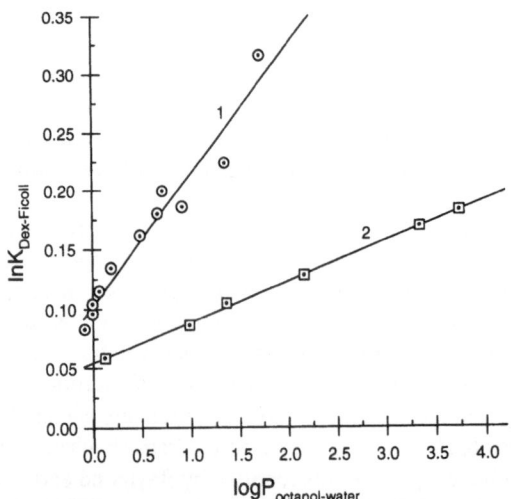

Figure 1. Relationships between partition coefficients of solutes in an aqueous Dextran-Ficoll two-phase system (containing 0.15 mole/kg NaCl in 0.01 mole/kg sodium phosphate buffer, pH 7.4) and side-chain contributions into log P for dinitrophenylated amino acids (1); and log P for morphine and its derivatives (2).

The data calculated from the results reported in references 14, 18, and 20 and plotted in Figure 1 indicate that there are linear relationships between ln K and log P for small organic molecules. As mentioned above, log P for a given series of solutes is known to usually be related to the biological potency of the solutes for conventional drugs. Hence, the logarithm of the partition coefficient of a biological solute in an aqueous two-phase system, ln K, may be expected to be related to the solute biological potency.

Opioid peptides and their analogs that interact with opiate receptors in much the same way as morphine and morphine-like conventional drugs were examined by Zaslavsky *et al.*[20] as endogenous biological solutes. The relative hydrophobicity of a number of peptides was measured by partitioning in aqueous Dextran-Ficoll two-phase systems containing different amounts of sodium phosphate buffer, pH 7.4, and NaCl.

The correlation between the relative hydrophobicity of the peptides and their affinity for [^{3}H]-naloxone binding sites in rat brain homogenate is shown in Figure 2. This correlation (curve 2) is described as:

$$\log (1/C_{50})_{RBR} = 6.88 + 20.3 \cdot \ln K - 51.4 \cdot (\ln K)^2 \tag{1}$$
$$N = 16; \ r^2 = 0.987; \ s = 0.044$$

where K is the partition coefficient of a peptide in the aqueous Dextran-Ficoll system of a fixed salt composition with ionic strength of 0.170 M and C_{50} is the peptide concentration producing 50% inhibition of the binding of [^{3}H]-naloxone to rat brain homogenate.

The activities of morphine, nalorphine, d-methadone, levorphanol, codeine, and naloxone in the rat brain receptors binding assay are plotted in Figure 2 (curve 1) versus the relative hydrophobicity of the drugs measured by the aqueous two-phase partitioning technique.

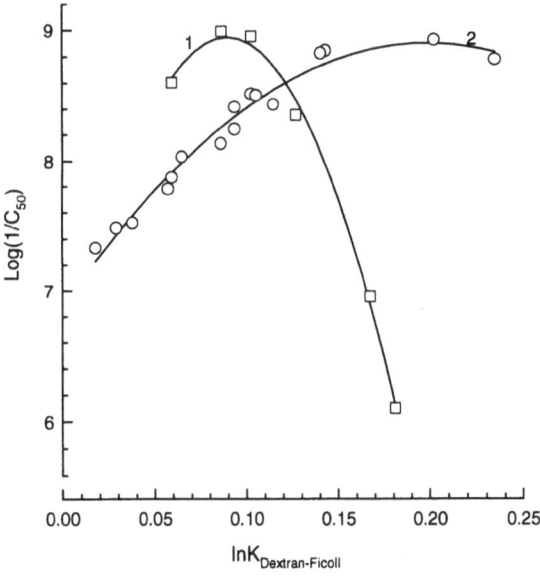

Figure 2. Relationships between the activity of morphine-like drugs (1) and opioid peptides (2) in rat brain receptor binding assay (expressed as log (1/C_{50})) and the relative hydrophobicity of compounds (expressed as ln K).

The relationship represented by curve 1 is described as:

$$\log (1/C_{50})_{RBR} = 6.23 + 60.5 \cdot \ln K - 338 \cdot (\ln K)^2 \qquad (2)$$
$$N = 6; \ r^2 = 0.998; \ s = 0.044$$

where ln K is the partition coefficient of a morphine-like drug in the aqueous Dextran-Ficoll two-phase system and C_{50} characterizes the drug binding to rat brain homogenate.

Equation (2) indicates that the "optimal" relative hydrophobicity, ln K_o, of the enkephalin-like peptide displaying the maximal potency in the rat brain homogenate assay is equivalent to that of 7.3 ± 0.3 CH$_2$ groups (see reference 6). The "optimal" value for opiates appear to be much lower, amounting to 3.3 ± 0.1 equivalent CH$_2$ groups. The difference between the relationships described by Equations (1) and (2) and the corresponding "optimal" ln K_o values for the peptides and morphine-like drugs is in line with that opioid peptides and opiates interact with different receptors. Qualitatively similar ln K-potency relationships were observed for opioid peptides in other bioassays as well.[6,20]

The first example of the ln K-potency relationship for proteins was obtained by Gulaeva et al.[21] for a number of lectins from different sources. Hemagglutinating activity toward rabbit red blood cells is plotted against the partition coefficients of the lectins in the aqueous Dextran-Ficoll two-phase system in Figure 3.

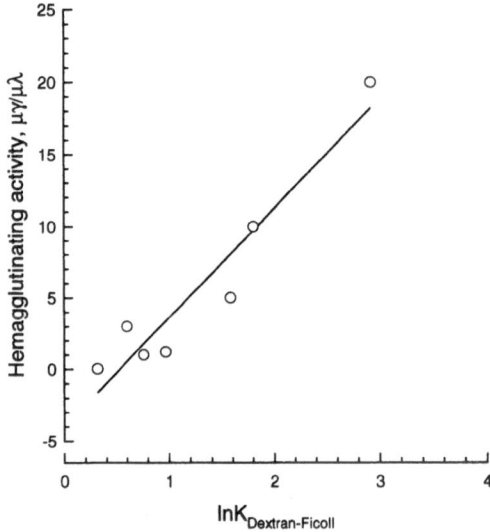

Figure 3. Relationship between the hemagglutinating activity and partition coefficients for various lectins in the aqueous Dextran-Ficoll two-phase system.

The relationships presented in Figure 3 may be described as:

$$C_{aggl} = -4.0 + 7.63 \cdot \ln K \qquad (3)$$
$$N = 7; \ r^2 = 0.955$$

where C_{aggl} is the limit concentration (in μg/mL) of lectin producing agglutination of rabbit erythrocytes and K is the partition coefficient of lectin in the aqueous Dextran-Ficoll two-phase system containing 0.15 mole/kg NaCl in 0.01 mole/kg sodium phosphate buffer, pH 7.4.

The relationship described by Equation (3) is, to my knowledge, the first example of a quantitative structure-activity relationship for proteins using the partition coefficient (relative hydrophobicity) as a general structural descriptor of a protein.

An additional relationship to be mentioned is the one established by Gulaeva *et al.*[22] between the partition coefficients of the 1:2 human serum albumin-drug complexes in the aqueous Dextran-Ficoll two-phase system and the drug half-life time in humans for sulfonamides. This relationship is shown in Figure 4.

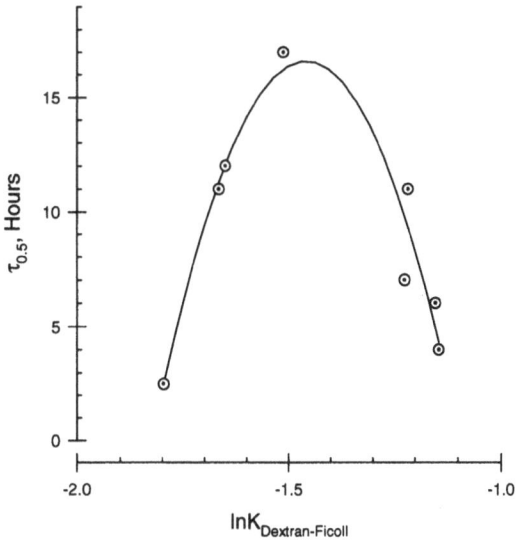

Figure 4. Relationship between the half-life time of sulfonamides in humans, $\tau_{0.5}$, and the partition coefficient of the albumin-sulfonamide 1:2 complex in the aqueous Dextran-Ficoll two-phase system.

The relationship shown in Figure 4 may be described as:

$$\tau_{0.5} = -247 - 361 \cdot \ln K - 124 \cdot (\ln K)^2 \qquad (4)$$
$$N = 8; \ r^2 = 0.971$$

where K is the partition coefficient of the human serum albumin-sulfonamide 1:2 complex in the aqueous Dextran-Ficoll two-phase system containing 0.15 mole/kg NaCl in 0.01 mole/kg sodium phosphate buffer, pH 7.4, and $\tau_{0.5}$ is the sulfonamide half-life time in humans.

The number of structure-potency correlations established for biological solutes so far is too limited to draw any general conclusion. Nevertheless, the existence of these relationships clearly indicates QSAR analysis to be possible for biological macromolecules with the aqueous two-phase partition coefficient as a general structure descriptor for conformationally flexible biological solutes, such as peptides, proteins, glycoproteins, etc.

The above data also indicate that the partition coefficient of a biological solute is related to the solute biological potency. This fact combined with the aforementioned high sensitivity of the solute partition behavior toward small changes in the solute structure, implies that the aqueous two-phase partition technique may be used as a bioanalytical method for characterization, quality control, and assessment of the lot-to-lot consistency of recombinant products.

To verify this possibility, partitioning of samples from ten different lots of the therapeutic recombinant human growth hormone preparation manufactured by Scientific-Industrial Association "Biotechnologia" (Moscow, Russia) was examined by Gulaeva et al.[23] in the aqueous Dextran-polyvinylpyrrolidone two-phase system containing an additive of sodium sulfate. The partition coefficient values of the samples were compared to that of the reference sample (2.58 ± 0.12). The partition coefficient K values for different samples varied from 2.48 to 4.07. No correlation between the K values and electrophoretic purity, monomer content, or results of the RIA (radioimmunoassay) analysis could be found. Only about half of the samples displayed partition behavior identical (within the experimental error limits) with that of the reference sample. A relationship between the partition coefficient value of a hormone sample and its biological potency was observed. The partition technique was found to be much more sensitive and more time- and cost-effective than other methodologies currently used for analysis of the therapeutic recombinant human growth hormone preparations.[24]

The above results indicate that the aqueous two-phase partition technique can be used as a simple, inexpensive, and highly sensitive method for analysis and quality control of recombinant products and other biological materials.

REFERENCES

1. P.Å. Albertsson. "Partition of Cell Particles and Macromolecules," 3rd. ed., Wiley, New York (1986).
2. "Partitioning in Aqueous Two-Phase Systems. Theory, Methods, Uses, and Applications to Biotechnology," H. Walter, D.E. Brooks, D. Fisher, eds., Academic Press, Orlando, Florida (1985).
3. W. Muller. "Liquid-Liquid Partition Chromatography of Biopolymers," GIT Verlag, Darmstadt (1988).
4. "Separations Using Aqueous Phase Systems. Applications in Cell Biology and Biotechnology," D. Fisher and I.A. Sutherland, eds., Plenum Press, New York (1989).
5. "Methods in Enzymology, Vol. 228, Aqueous Two-Phase Systems," H. Walter and G. Johansson, eds., Academic Press, Orlando, Florida (1994).
6. B.Y. Zaslavsky. "Partitioning in Aqueous Two-Phase Systems: Physical Chemistry and Bioanalytical Applications," Marcel Dekker, New York (1994).
7. F.D. Raymond, D.W. Moss, and D. Fisher, Phase partitioning detects differences between phospholipase-released forms of alkaline phosphatase - a GPI-linked protein, *Biochim. Biophys. Acta* 1156:117 (1993).
8. A.D. Diamond, K. Yu, and J.T. Hsu, The effect of amino acid sequence on peptide and protein partitioning in aqueous two-phase systems, *in*: "Protein Purification: From Molecular Mechanisms to Large-Scale Processes," M.R. Ladisch, R.C. Willson, C.C. Painton, and S.E. Builder, eds., ACS Symposium Series, Vol. 427, pp. 52-65, American Chemical Society, Washington, D.C. (1990).
9. V.V. Mozhaev, V.A. Siksnis, N.S. Melik-Nubarov, N.Z. Galkantaite, G.J. Denis, E.P. Butkus, B.Y. Zaslavsky, N.M. Mestechkina, and K. Martinek, Protein stabilization via hydrophilization. Covalent modification of trypsin and α-chymotrypsin, *Eur. J. Biochem.* 173:147 (1988).
10. S.-O. Enfors, K. Köhler, and A. Veide, Combined use of extraction and genetic engineering for protein purification: recovery of β-galactosidase fused proteins, *Bioseparation* 1:305 (1990).
11. J.R. Luther and C.E. Glatz, Genetically engineered charge modifications to enhance protein separation in aqueous two-phase systems: electrochemical partitioning, *Biotechnol. Bioeng.* 44:147 (1994).
12. H. Tanaka, R. Kuboi, and I. Komasawa, The effect of hydrochloric acid on hydrophobicity and partition of protein in aqueous two-phase systems, *J. Chem. Eng. Jpn.* 24:661 (1991).
13. R. Kuboi, K. Yano, H. Tanaka, and I. Komasawa, Evaluation of surface hydrophobicities during refolding process of carbonic anhydrase using aqueous two-phase partitioning systems, *J. Chem. Eng. Jpn.* 26:286 (1993).
14. B.Y. Zaslavsky, N.M. Mestechkina, L.M. Miheeva, and S.V. Rogozhin, Measurements of relative hydrophobicity of amino acid side-chains by partition in an aqueous two-phase polymeric system: hydrophobicity scale for non-polar and ionogenic side-chains, *J. Chromatogr.* 240:21 (1982).

15. B.Y. Zaslavsky and E.A. Masimov, Methods of analysis of the relative hydrophobicity of biological solutes, *Topics Curr. Chem.* 146:171 (1988).

16. B.Y. Zaslavsky, A.A. Borovskaya, N.D. Gulaeva, and L.M. Miheeva, Partitioning in aqueous two-phase systems as a method of estimation of the relative hydrophobicity of solutes, *J. Chem. Soc. Faraday Trans.* 87:137 (1991).

17. P.A. Rebinder, Hydrophilicity and hydrophobicity, *in*: "Kratkaya Khimicheskaya Encyclopedia," Vol.1, p. 937, Sovetskaya Encyclopedia, Moscow (1961).

18. C. Hansch and A. Leo. "Substituent Constants for Correlation Analysis in Chemistry and Biology," Academic Press, New York (1979).

19. R. Franke. "Theoretical Drug Design Methods", Elsevier, Amsterdam (1984).

20. B.Y. Zaslavsky, N.M. Mestechkina, L.M. Miheeva, S.V. Rogozhin, G.Y. Bakalkin, G.G. Rjazhsky, E.V. Chetverina, A.A. Asmuko, J.D. Bespalova, N.V. Korobov, and O.N. Chichenkov, Correlation of hydrophobic character of opioid peptides with their biological activity measured in various bioassay systems, *Biochem. Pharmacol.* 31:3757 (1982).

21. N.D. Gulaeva, I.A. Yamskov, and B.Y. Zaslavsky, unpublished data (1990).

22. N.D. Gulaeva and B.Y. Zaslavsky, unpublished data (1990).

23. N.D. Gulaeva, M.A. Chlenov, B.Y. Zaslavsky, unpublished data (1989).

24. A.F. Bristow and S.L. Jeffcoate, Analysis of therapeutic growth hormone preparations: report of an interlaboratory collaborative study on growth hormone assay methodologies, *Biologicals* 20:221 (1992).

INDEX